LAMINATION - THEORY AND APPLICATION

Edited by **Charles A. Osheku**

Lamination - Theory and Application
http://dx.doi.org/10.5772/65194
Edited by Charles A. Osheku

Contributors

Charles Attah Osheku, Baoxi Liu, Lujun Huang, Lin Geng, Fuxing Yin, Marina Rakocevic, Dionysios E. Mouzakis, Mizuho Shibata, Norimitsu Sakagami

Published by InTech
Janeza Trdine 9, 51000 Rijeka, Croatia

Publishing Process Manager Mirena Calmic
Technical Editor SPi Global
Cover InTech Design team

First published March, 2018
Printed in Croatia
Legal deposit, Croatia: National and University Library in Zagreb

Additional hard copies can be obtained from orders@intechopen.com

Lamination - Theory and Application, Edited by Charles A. Osheku
p. cm.
Print ISBN 978-953-51-3925-6
Online ISBN 978-953-51-3926-3

For EU product safety concerns:
IN TECH d.o.o., Prolaz Marije Krucifikse Kozulić 3, 51000 Rijeka, Croatia,
info@intechopen.com or visit our website at intechopen.com.

Meet the editor

Dr. Charles Attah Osheku is the current director of the Centre for Space Transport and Propulsion, Epe, Lagos, an activity centre of the National Space Research and Development Agency, Nigeria. He obtained his BSc degree in Mechanical Engineering from the Obafemi Awolowo University and his MSc and PhD degrees in Mechanical Engineering (Engineering Mechanics) from the University of Lagos. Dr. Osheku is a research expert in the fields of rocket and missile solid fuel physics; solid rocket engine design; structural-, flow- and combustion-induced vibrations; aircraft noise modelling, and dissipation and vibration of laminated structures. Dr. Osheku has 66 scientific papers to his credit, published in reputable international journals and ASME International Offshore Mechanics Conferences. He is a registered engineer and a professional member of ASME and AIAA.

Contents

Preface

The field of lamination has developed significantly over the past 5000 years. Nowadays, we have a humongous array of structures and technological systems where composite laminates are applied. These range from aviation, aerospace/space to civil and marine structures.

From the viewpoint of structural mechanics, an interface slip motion between two laminated structures, such as beam plate and plate in the presence of dry friction, can be utilized for slip damping systems. By scientific definition, slip damping is a mechanism exploited for dissipating noise and vibration energy in machine structures and systems. Researchers have developed several mathematical models for noise dissipation, minimization and complete vibration isolation laminated mechanisms. The purpose of this book is to describe new concepts of producing laminated structures and possible modern engineering applications.

The introductory chapter of this book offers new engine design concepts on how laminated enclosures can be useful in aerospace, machine and ocean structures. Chapter 2 presents research results on how higher strength can be achieved in titanium matrix composites by adjusting the multi-scale hierarchical structure. In Chapter 3, the proposed higher-order zigzag theory is expected to ease the determination of deflections and stresses in composite sandwich laminate analysis. In Chapter 4, the dynamic modelling of a serial link robot laminated with plastic film to improve waterproofing and dust-proofing of serial link robots is the main focus. Chapter 5 emphasizes the importance of additive 3D manufacturing techniques and their robust applications in aeronautical and defence structural systems.

A number of technological applications are possible in the areas of machine structures, power plants, ship turbines, military jets and commercial aircrafts with suitable laminated materials in conjunction with intelligent manufacturing systems to produce high-tech devices/systems that are more comfortable for use by humans especially in the modern-day world with advanced 3D manufacturing techniques.

Charles A. Osheku, PhD
Director/Chief Executive Officer
Centre for Space Transport and Propulsion
National Space Research and Development Agency
Federal Ministry of Science and Technology
LASU, Epe Campus, Epe, Lagos, Nigeria

Section 1

Introduction

Chapter 1

Introductory Chapter: Laminations - Theory and Applications

Charles Attah Osheku

Additional information is available at the end of the chapter

http://dx.doi.org/10.5772/intechopen.74457

1. Introduction

Historically, the earliest use of laminated materials was by the Mesopotamians around 3400 BC. Similarly, plywood laminates from wood strips glued together at different angles were constructed for structural applications. Nevertheless, for several centuries, primitive lamination methods were used for the construction of bows and canoes. Specifically, the invention of Bakelite in 1907 heralded a new era for composite and laminated material applications. For instance, at the inception of the aircraft by the Orville and Wilbur Wright brothers, it was made with wood and fabrics. The need for better aerodynamic performance and efficiency supported by advances in aerospace engineering led to the use of lighter materials with strength that matches that of steel. The development of this field has progressed significantly to the point of the ubiquitous availability of laminates in materials for aviation, civil structures and automotive uses. In particular, carbon fiber reinforced plastics are in use in modern aircraft aerospace systems. In modern engineering practices, the use of layered sandwich structural flat plate members is increasing steadily in aerospace, civil, mechanical and offshore structures due to their high specific strength and stiffness. Such structural applications have significantly led to reduction in vulnerability of warships to blasts, ballistics, bombs and fire attacks and to enhance superior resistance to fatigue crack propagation, impact damage and local buckling.

From the viewpoint of structural mechanics, an interface slip motion between two laminated structures, such as beam, beam plate and plate in the presence of dry friction can be utilized for slip damping systems. By scientific definition, slip damping is a mechanism exploited for dissipating noise and vibration energy in machine structures and systems. There are several engineering procedures to effect or simulate such a damping phenomenon. For example, the introduction of constrained, unconstrained and even viscoelastic layers has been very helpful in this respect. One good technique is layered construction made possible by externally applied pressure at the interface of two structural members. An arrangement of this nature could either be jointed or fastened by appropriate bolting methods. Within that concept, interface pressure profiles must assume a significant role subject to interfacial slip motion to waste or dissipate energy from the induced vibration arising from any form of excitation.

Within this context, researchers have developed a number of interesting mathematical models to derive the advantages discussed above for noise dissipation, minimization or complete vibration isolation. For instance, in the paper published by Osheku et al. [1], they employed a mathematical model to discuss the effect of structural vibration on the propagation of acoustic pressure waves through a cantilevered 3D laminated beam-plate enclosure. On the other hand, slip damping with heterogeneous sandwich composite viscoelastic beam-plate smart systems as a model for dissipation of vibration and active noise control mechanism in ship and floating structures was studied effectively by Olunloyo and Osheku in [2]. A scientific discussion of interfacial slip through layered laminates is described as shown in **Figure 1**.

As a demonstration of the application of laminations to noise reduction, **Figure 2** shows the problem geometry subject to governing equations as derived in [1]:

$$\nabla^2 P - \frac{\partial^2 P}{c_1^2 \partial t^2} - \frac{2u_x \partial^2 P}{c_1^2 \partial t \partial x} - \frac{u_x^2}{c_1^2}\frac{\partial^2 P}{\partial x^2} = -\frac{\partial^2 W_1}{\rho \partial t^2} \tag{1}$$

while the governing equations for each vibrating boundary are namely:

For the case Ω_1,

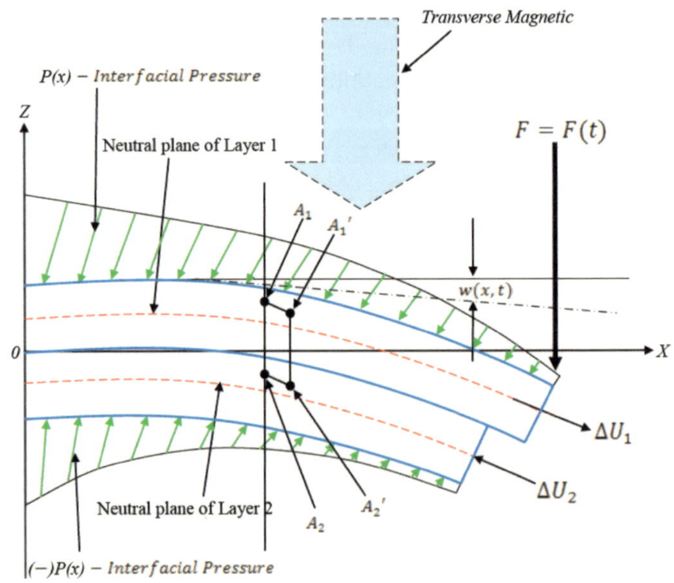

Figure 1. Mechanism of interfacial slip geometry.

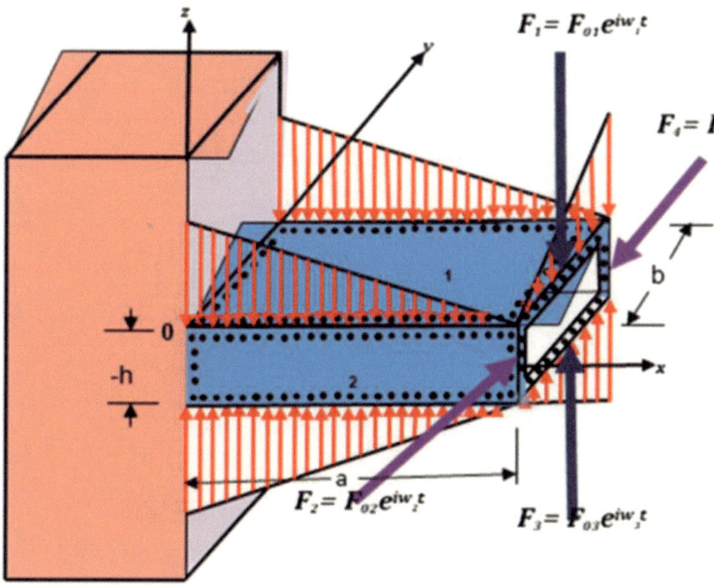

Figure 2. Problem geometry of 3D composite structure.

$$D\left(\frac{\partial^4 W_1}{\partial x^4}+2\frac{\partial^4 W_1}{\partial x^2 \partial y^2}+\frac{\partial^4 W_1}{\partial y^4}\right)+\rho_1 h\frac{\partial^2 W_1}{\partial t^2}-\left(\frac{\mu H_2}{2}\frac{\partial p(x,0)}{\partial x}+\frac{\mu H_2}{2}\frac{\partial p(0,y)}{\partial y}\right)=0 \tag{2}$$

For the case Ω_2,

$$\frac{\partial^2 W_2}{\partial x^2}+\beta_2\frac{\partial^2 W_2}{\partial t^2}=\alpha_2\frac{\partial P(x,0)}{\partial x} \tag{3}$$

For the case Ω_3,

$$D\left(\frac{\partial^4 W_3}{\partial x^4}+2\frac{\partial^4 W_3}{\partial x^2 \partial y^2}+\frac{\partial^4 W_3}{\partial y^4}\right)+\rho_1 h\frac{\partial^2 W_3}{\partial t^2}-\left(\frac{\mu H_2}{2}\frac{\partial p(x,0)}{\partial x}+\frac{\mu H_2}{2}\frac{\partial p(0,y)}{\partial y}\right)=0 \tag{4}$$

For the case Ω_4,

$$\frac{\partial^2 W_4}{\partial x^2}+\beta_2\frac{\partial^2 W_4}{\partial t^2}=\alpha_4\frac{\partial P(x,0)}{\partial x} \tag{5}$$

$$\forall \alpha_1=\alpha_2=\alpha_3=\alpha_4=\frac{6\mu}{Eh^2};\ \beta_1=\beta_2=\beta_3=\beta_4=\frac{\rho bh}{EI} \tag{6}$$

The boundary stresses through the domains Ω_1, Ω_2 and Ω_3 are evaluated from the following equations in Fourier finite double sine transforms plane, viz.:

$$\tilde{p}^{F_yF_z}(0,\lambda_m,\lambda_k,s)=\int_0^b\int_0^c[\mu P_{av}]\sin\left(\frac{m\pi y}{b}\right)\sin\left(\frac{k\pi z}{c}\right)dydz,$$

$$\tilde{p}^{F_yF_z}(a,\lambda_m,\lambda_k,s)=\int_0^b\int_0^c[\mu P_{av}]\sin\left(\frac{m\pi y}{b}\right)\sin\left(\frac{k\pi z}{c}\right)dydz \tag{7}$$

$$\tilde{p}^{F_xF_y}(\lambda_n,\lambda_m,0,s)=\int_0^b\int_0^c\left[\frac{P_{av}}{s}\right]\sin\left(\frac{n\pi x}{a}\right)\sin\left(\frac{m\pi y}{b}\right)dxdy$$

$$\tilde{p}^{F_xF_y}(\lambda_n,\lambda_m,c,s)=-\int_0^b\int_0^a\left[\frac{P_{av}}{s}\right]\sin\left(\frac{m\pi x}{a}\right)\sin\left(\frac{k\pi y}{b}\right)dxdy, \tag{8}$$

$$\tilde{p}^{F_xF_z}(\lambda_n,b,\lambda_k,s)=\int_0^a\int_0^c\left[\frac{P_{av}}{s}\right]\sin\left(\frac{n\pi x}{a}\right)\sin\left(\frac{k\pi z}{c}\right)dxdz, \tag{9}$$

$$\tilde{p}^{F_xF_z}(\lambda_n,0,\lambda_k,s)=-\int_0^a\int_0^c\left[\frac{P_{av}}{s}\right]\sin\left(\frac{n\pi x}{a}\right)\sin\left(\frac{k\pi z}{c}\right)dxdz, \tag{10}$$

subject to interfacial slip mechanism as modulated by the pressure profile as discussed in the paper. In the meantime, when such a slip mechanism is to be examined under the influence of an electromagnetic field, the details of such a problem were discussed in Osheku [3].

The underlying principle is such that during bending, each half of the sandwich elastic in transverse magnetic field has its neutral plane that does not necessarily coincide with the geometric mid-plane through the interface because of the frictional stresses. These are located at $z=\alpha_1(x)\frac{h}{2}$ and $z=\alpha_2(x)\frac{h}{2}$ on layers (1) and (2) respectively, where $\alpha_1(x)$ and $\alpha_2(x)$ are generalized functions of x. The use of Taylor series approximation leads to the following expressions for the microscopic slip or displacements of the two adjacent opposite points A_1 and A_2 to the corresponding ${A_1}^1$ and ${A_2}^1$ namely:

$$\begin{aligned}\Delta U_1(x,z,t)=\Delta U_1(0,0+,t)+\left(z-\alpha_1(x)\frac{h}{2}\right)\frac{\partial W(x,t)}{\partial x}+\frac{1}{2}\left(z-\alpha_1(x)\frac{h}{2}\right)^2\frac{\partial^2 W(x,t)}{\partial x^2}\\+\frac{1}{6}\left(z-\alpha_1(x)\frac{h}{2}\right)^3\frac{\partial^3 W(x,t)}{\partial x^3}\end{aligned} \tag{11}$$

and

$$\begin{aligned}\Delta U_2(x,z,t)=\Delta U_2(0,0-,t)-\left(z+\alpha_2(x)\frac{h}{2}\right)\frac{\partial W(x,t)}{\partial x}-\frac{1}{2}\left(z+\alpha_1(x)\frac{h}{2}\right)^2\frac{\partial^2 W(x,t)}{\partial x^2}\\-\frac{1}{6}\left(z+\alpha_1(x)\frac{h}{2}\right)^3\frac{\partial^3 W(x,t)}{\partial x^3}\end{aligned} \tag{12}$$

In the meantime, a first-order approximation follows as:

$$\Delta U_1(x,z,t) = \Delta U_1(0,0+,t) + \left(z - \alpha_1(x)\frac{h}{2}\right)\frac{\partial W(x,t)}{\partial x} \tag{13}$$

$$\Delta U_2(x,z,t) = \Delta U_2(0,0-,t) - \left(z + \alpha_2(x)\frac{h}{2}\right)\frac{\partial W(x,t)}{\partial x} \tag{14}$$

Here, $\Delta U_1(0,0+,t)$ and $\Delta U_2(0,0-,t)$ must be zero at the fixed end.

$$\Delta U(x,z,t) = \Delta U_1(x,z,t) - \Delta U_2(x,z,t) \tag{15}$$

The associated interfacial slip motion is then given by the equation below which on following Goodman and Klumpp [4] becomes:

$$\Delta U(x,0) = E^{-1}\int_0^x \left[(\sigma_x)_1(\xi,0+,t) - (\sigma_x)_2(\xi,0+,t)\right]d\xi \tag{16}$$

where ξ is a dummy axial spatial variable of integration across the interface and 0+, 0− denote the origin of the transverse spatial variable for each layer; while subscripts 1 and 2 refer to the upper and lower laminates.

For the contrived geometry, the derived corresponding spatial bending stresses are namely:

$$(\sigma_x)_1(x,z,t) = -\frac{E}{2}(2z-h)\frac{\partial^2 W(x,t)}{\partial x^2} + \frac{B_0^2}{\mu_0}\left(1-\frac{\mu_0}{\mu_m}\right)[W(x,t)-W(L,t)] + \mu P_0(1+\varepsilon\overline{x})\frac{(x-L)}{h} \tag{17}$$

and

$$(\sigma_x)_2(x,z,t) = -\frac{E}{2}(2z+h)\frac{\partial^2 W(x,t)}{\partial x^2} + \frac{B_0^2}{\mu_0}\left(1-\frac{\mu_0}{\mu_m}\right)[W(x,t)-W(L,t)] - \mu P_0(1-\varepsilon\overline{x})\frac{(x-L)}{h} \tag{18}$$

This gives Eq. (11) as:

$$\Delta U(\overline{x}) = \int_0^x \left[\frac{\partial^2 \overline{W}(\overline{x},\tau)}{\partial \overline{\xi}^2} + \overline{B}_0^2\left(1-\frac{\mu_0}{\mu_m}\right)\left[\overline{W}(\overline{\xi},\tau) - \overline{W}(1,\tau)\right] + 2\mu P_0\left(1+\varepsilon\overline{\xi}\right)\left(\overline{\xi}-1\right)\right]d\overline{\xi} \tag{19}$$

and on introducing the nondimensionalized parameters, viz.:

$$\Delta U = \frac{\Delta U(\overline{x})Ebh^2}{L^2F}; \overline{P}_0 = \frac{P_0}{F/bh}; t = \frac{2\pi\tau}{\omega_0} \tag{20}$$

For the static case, the governing differential equation is described in [2] as

$$EI\frac{d^4W}{dx^4}+\frac{bh}{\mu_0}\left(1-\frac{\mu_0}{\mu_m}\right)B_o^2\frac{d^2W}{dx^2}=\frac{1}{2}bh\mu\frac{dP}{dx} \tag{21}$$

subject to the form computed as in [3] as:

$$\overline{P}_{av}=\overline{P}_o\int_0^1(1+\varepsilon\overline{x})d\overline{x}=\overline{P}_o\left(1+\frac{\varepsilon}{2}\right) \tag{22}$$

In the meantime, several derivatives of this analysis can be conjured for comparative analysis with literature.

For the case of a layer of sandwich homogenous magnetoelastic beam-plate of thickness h with non-uniform pressure at the interface, the formulated equation governing the vibration takes the form:

$$D\frac{\partial^4W}{\partial x^4}+\frac{h}{\mu_0}\left(1-\frac{\mu_0}{\mu_m}\right)B_0{}^2\frac{\partial^2W}{\partial x^2}-\frac{\sigma_c h^3B_0{}^2}{12}\frac{\partial^3W}{\partial t\partial x^2}+\rho h\frac{\partial^2W}{\partial t^2}=\frac{1}{2}h\mu\frac{\partial P}{\partial x} \tag{23}$$

$$\forall D=\frac{Eh^3}{12(1-v^2)} \tag{24}$$

so that for the case corresponding to uniform interfacial pressure, Eq. (23) reduces to the following:

$$D\frac{\partial^4W}{\partial x^4}+\frac{h}{\mu_0}\left(1-\frac{\mu_0}{\mu_m}\right)B_0{}^2\frac{\partial^2W}{\partial x^2}-\frac{\sigma_c h^3B_0{}^2}{12}\frac{\partial^3W}{\partial t\partial x^2}+\rho h\frac{\partial^2W}{\partial t^2}=0 \tag{25}$$

Nonetheless, a theoretical investigation by [5] on the proposed uses of lamination theory via slip damping was hinged on the following pictorial representation in **Figure 3**.

Figure 3 illustrates a slip damping theory and its application in ship hull subject to Eq. (26). Detailed solution methods of this boundary value mathematical physics problem are outlined in [5]. The intent here is to demonstrate possible means to enhance moving ocean craft stability subject to turbulent flow conditions in **Figure 4**. Meanwhile, the governing equation is depicted as follows,

$$\left(\frac{c_1+c_2}{4}\right)\frac{\partial^5W}{\partial t\partial x^4}+\frac{\partial^4W}{\partial x^4}+\frac{(\beta_1\mathcal{A}+\beta_2\mathcal{B}+\beta_3+\beta_4)}{4}\frac{\partial^2W}{\partial t^2}=\frac{(\alpha_1\mathcal{A}+\alpha_2\mathcal{B})}{4}\frac{\partial P(x,0)}{\partial x}+\frac{C_a}{4}\frac{\partial^2V_a}{\partial x^2} \tag{26}$$

where $\mathcal{A}$ denotes $1-v_1^2$ and β denotes $1-v_2^2$ and the following parameters have been defined viz.:

$$\alpha_1=\frac{6\mu}{E_{eq1}h_{eq1}^2},\alpha_2=\frac{6\mu}{E_{eq2}h_{eq2}^2} \tag{27}$$

$$\beta_1=\frac{\rho_1bh_1}{E_1I_1},\beta_2=\frac{\rho_2bh_2}{E_2I_2},\beta_3=\frac{\rho_abh_a}{E_aI_{a1}},\beta_4=\frac{\rho_sbh_s}{E_sI_s} \tag{28}$$

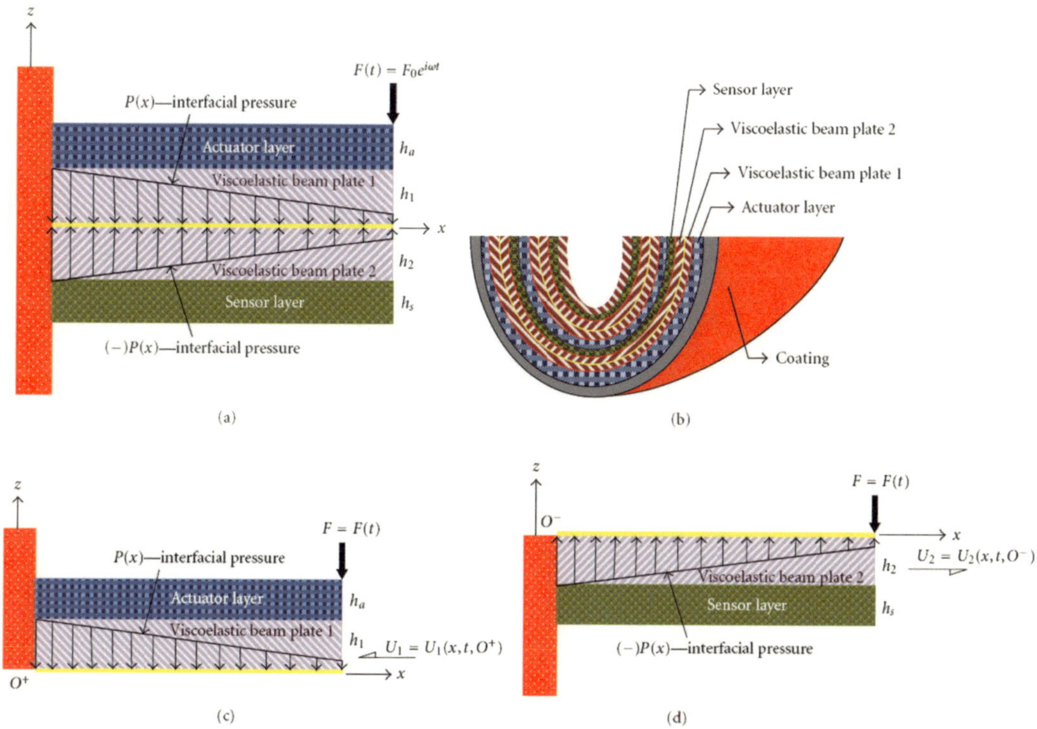

Figure 3. (a) Preslip geometry for the composite sandwich structure under dynamic load. (b) Layering cross section of composite structure.

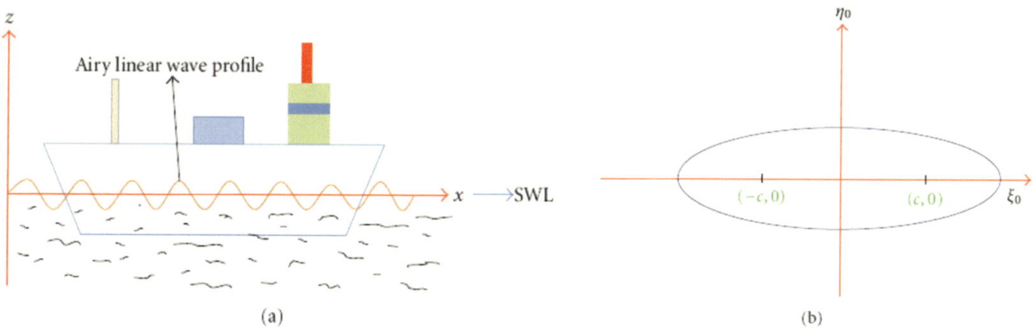

Figure 4. (a) A mathematical model of ship/Floating Production Storage and Offloading vessel (FPSO) in ocean environment. (b) An elliptical model representation of a ship/FPSO bottom hull configuration.

$$h_{eq1} = (h_1 + h_a), h_{eq1} = (h_a + h_s), C_a = \frac{1}{2} E_a d_{31} b(h_a + h_s) \tag{29}$$

By utilizing the closed form expression of the interfacial slip motion, the energy dissipation quantification can be evaluated from ongoing equation

$$D = 4\mu b \int_0^{\pi/2\omega} \int_0^L P(x)\Delta u(x,t) f dx\, dt \tag{30}$$

and nondimensionalized as:

$$\overline{D} = 4\mu \int_0^{1/4} \int_0^1 \overline{P}_{av} \Delta \overline{u} d\overline{x} d\tau, \tag{31}$$

where $\overline{D} = D(\overline{x}, \tau) Ebh^3 / L^3 F_0^2$

For the purpose of clarity, some nomenclatures are described. Readers are advised to read the papers in the references for detailed description of symbols in the equations. Theoretical simulated studies of the foregoing based on the generalized energy dissipation equations and possible scientific applications and engineering design purposes are well highlighted in these papers.

2. Proposed applications

A number of technological applications are possible in the areas of machine structures, power plants, ship turbines, military jets and commercial aircrafts with suitable laminated materials in conjunction with intelligent manufacturing systems to produce systems that are more comfortable for use by humans especially in the modern day world with advanced 3D manufacturing techniques. Arising from research results based on lamination theory and applications, the following modern noise dissipation and vibration isolation systems can be incorporated in new aircraft engines and power generating turbo machines to enhance their operational stability are itemized in the following diagrams (**Figures 5–11**).

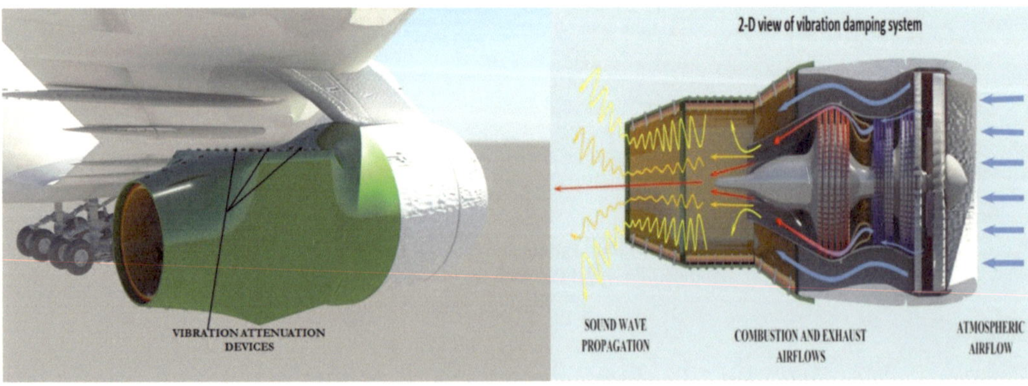

(a) (b)

Figure 5. (a) A commercial airliner with proposed vibration attenuation devices. (b) Proposed 2D cross section of vibration damping system.

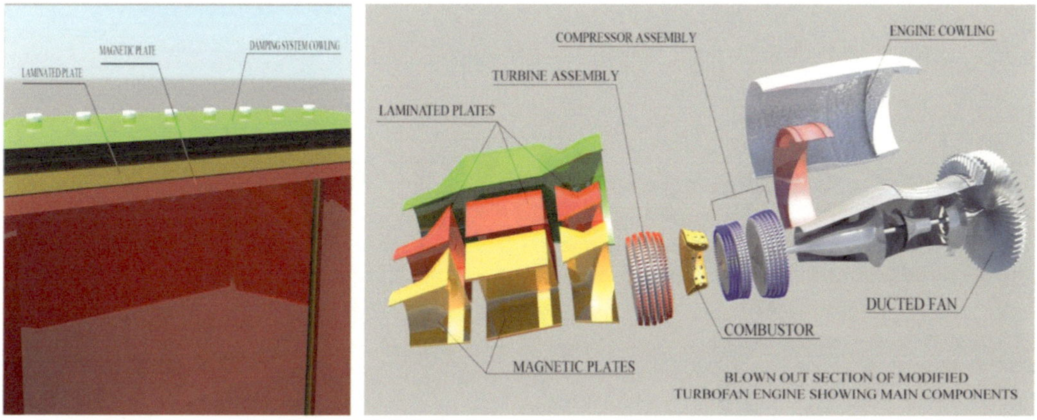

(a) (b)

Figure 6. (a) Sectional view through a modified aircraft engine. (b) Blown-out section of modified new jet engine concept showing laminated vibration damping mechanism.

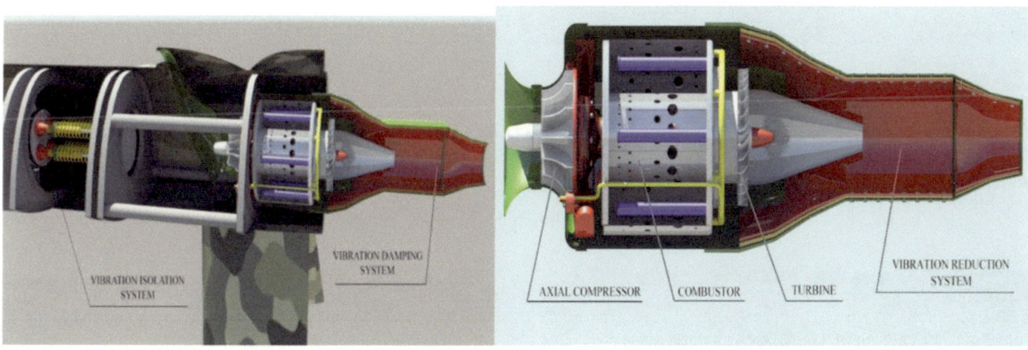

(a) (b)

Figure 7. (a) Proposed application of lamination for vibration damping and isolating system. (b) Modified axially compressing turbo jet aircraft engines.

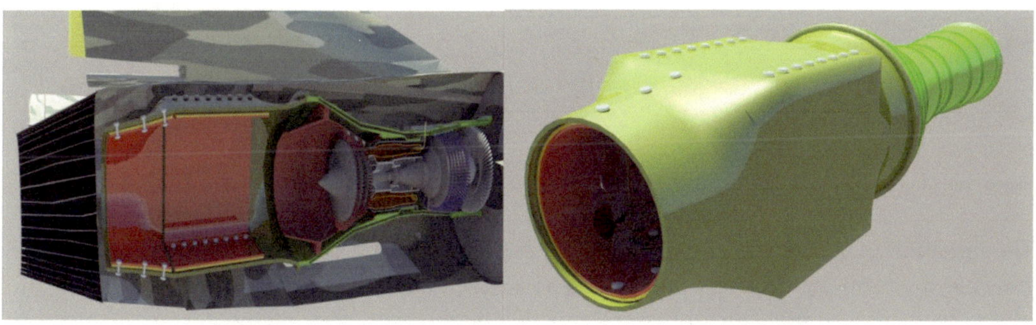

(a) (b)

Figure 8. (a) 3D section of jet fighter engine showing modified engine. (b) 3D view of jet fighter modified engine.

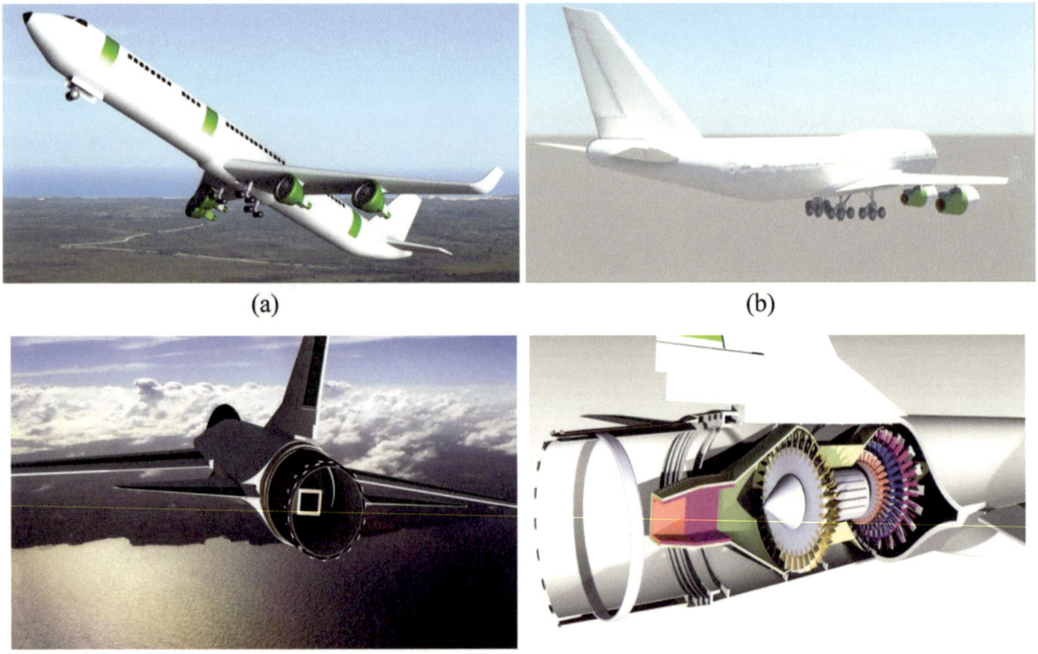

(a) (b) (c) (d)

Figure 9. (a, b) Proposed civilian aircraft with modified engines fitted with laminated vibration attenuation devices. (c) Proposed military jet fighter with modified engine fitted with laminated devices. (d) Proposed aircraft engine ducting with laminated devices.

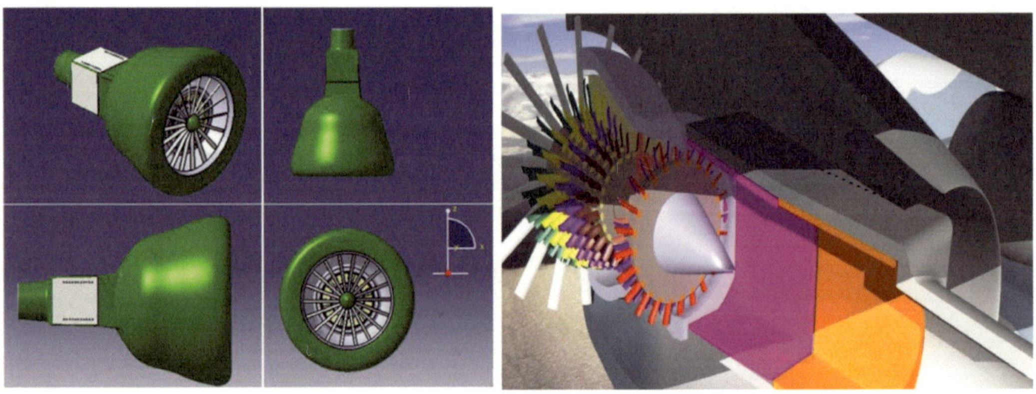

Figure 10. Jet engine vibration isolation system with laminated devices.

The intention of this book is to describe new concepts of producing laminated structures and possible modern engineering applications as demonstrated in the introductory chapter and other chapters. The introductory chapter of this book offers new engine design concepts on how laminated enclosures can be useful in machine, ocean and aerospace structures. The thrust is to showcase how the world can benefit from these innovative concepts.

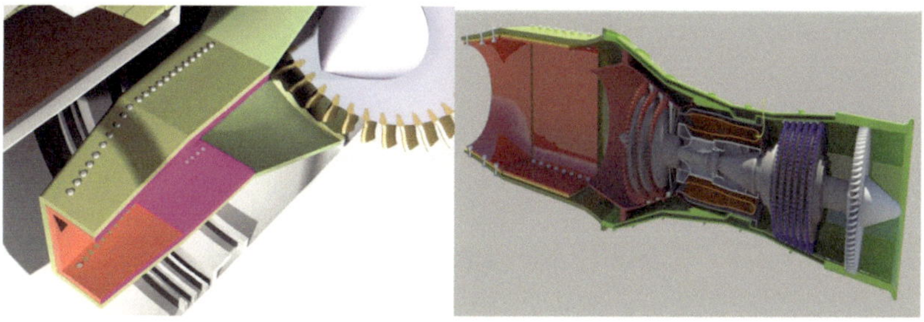

(a) (b)

Figure 11. (a, b) Proposed 3D view of a modified turbojet engine with laminated enclosure.

In Chapter 1, the focus is on the study of multiscale hierarchical structure and laminated strengthening and toughening mechanisms. Here, the authors emphasize how higher strength can be achieved in titanium matrix composites by adjusting the multiscale hierarchical structure. The chapter further deduced how the elastic properties and yield strength of laminated composites can be modulated through the "rule of average." It is also discussed in this chapter how impossible it is to predict fracture elongation and toughness via this rule. In addition, the authors pointed out that fracture elongations of laminated composites are closely tied to strain indexing exponent and strain rate parameter.

In Chapter 2, large deflection analysis of laminated composite plates using a higher order zigzag theory is discussed. This proposed refined theory is expected to ease the determination of deflections and stresses in composite sandwich laminate analysis. The presented model which incorporated an efficient C_0 plate finite element (FE) was shown to accurately calculate the deflections as well as stresses for different geometries of composite and sandwich laminates. The results obtained agree with existing exact 3D elasticity solutions for thin, moderately thick composite and sandwich laminates.

In Chapter 3, the dynamic modeling of a serial link robot laminated with plastic film to improve waterproofing and dustproofing of serial link robots is the main focus. It also discusses how to improve lubrication between the links and the film through an insulating fluid that was encapsulated in the plastic film by employing detailed appropriate mathematical analysis. Dynamic performance of the laminated body was validated for flexural rigidities. Through detailed mathematical analysis, the dynamic performance of the laminated body was confirmed for different flexural rigidities.

In Chapter 4, advanced technologies in manufacturing 3D layered structures for defense and aerospace systems is the main focus of the study. It emphasizes the importance of additive manufacturing techniques and their robust applications in aeronautical and defense structural systems. This study further reveals that not all advanced materials and alloys can be automatically layered by rapid prototyping system or machine. It also highlighted how efforts are underway to apply automated layering technology in many materials with potential applications in nowadays plastics and reinforced polymers for ease of manufacturing 3D parts. The discussion in this chapter presented a review of additive manufacturing history and the potential advantages the proposed method would offer.

Nomenclature

b	width of laminated beam
B_0	magnetic flux density
d/dx	differential operator
E	electric field intensity
E	modulus of rigidity
F	applied end force amplitude
h	depth of laminated beam
H	magnetic field intensity
I	moment of inertia
L	length of laminated beam
P	clamping pressure at the interface of the laminated beams
t	time coordinate
u, v	velocity
$\dot{u}, \dot{v}$	acceleration
U_1	displacement of the lower laminate
U_2	displacement of the upper laminate
W	dynamic response
W^F	dynamic response in Laplace transform plane
$\tilde{W}$	dynamic response in Fourier transform plane
$\tilde{W}^F$	dynamic response in Fourier-Laplace transform plane
x	space coordinate along the beam interface
z	space coordinate perpendicular to the beam interface
ε	pressure gradient
μ	dry friction coefficient
μ_m	permeability of the medium
μ_r	relative permeability of the medium
χ	normalized magnetic field intensity
ε_{x1}	axial strain in layer-1

ε_{x2} axial strain in layer-2

γ_{xz} angular strain

ρ density of laminate material

τ_{xz} shear stress at the interface of the laminates

$(\sigma_x)_1$ bending stress at the upper half of the laminates

$(\sigma_x)_2$ bending stress at the lower half of the laminates

ξ dummy variable

υ Poisson's ratio

Author details

Charles Attah Osheku

Address all correspondence to: charlesosheku2002@yahoo.com

Centre for Space Transport and Propulsion, National Space Research and Development Agency, Federal Ministry of Science and Technology, LASU Epe Campus, Epe, Lagos, Nigeria

References

[1] Osheku CA, Olunloyo VOS, Damisa O, Akano TT. Acoustic pressure waves in vibrating 3-D laminated beam-plate enclosures. Advances in Acoustics and Vibration. 2009;**2009**, Article ID 853407:14

[2] Olunloyo VOS, Osheku CA. On vibration and noise dissipation in ship and FPSO structures with smart systems. International Scholarly Research Network: Mechanical Engineering. 2012;**2012**, Article ID 127238:19

[3] Osheku CA. Mechanics of static slip and energy dissipation in sandwich structures: Case of homogeneous elastic beams in transverse magnetic fields. International Scholarly Research Network: Mechanical Engineering. 2012;**2012**, Article ID 372019:23

[4] Goodman LE, Klumpp JH. Analysis of slip damping with reference to turbine blade vibration. Journal of Applied Mechanics. 1956;**23**:421

[5] Olunloyo VOS, Osheku CA. On vibration and noise dissipation in ship and FPSO structures with smart systems. International Scholarly Research Network: Mechanical Engineering. 2012;**2012**, Article ID 127238:19

Section 2

Theory of Lamination

Chapter 2

Multiscale Hierarchical Structure and Laminated Strengthening and Toughening Mechanisms

Baoxi Liu, Lujun Huang, Lin Geng and Fuxing Yin

Additional information is available at the end of the chapter

http://dx.doi.org/10.5772/intechopen.69976

Abstract

Metal matrix composites with multiscale hierarchical structure and laminated structure have been developed to provide a novel route to achieve high strength, toughness and ductility. In this chapter, a lot of scientific research has been carried out in the preparation, processing, properties and application of metal matrix composite. Many toughening mechanisms and fracture behavior of composites with multiscale hierarchical structure and laminated structure are overviewed. It is revealed that elastic property and yield strength of laminated composites follow the "rule of average." However, the estimation of fracture elongation and fracture toughness is complex, which is inconsistent with the "rule of average." The fracture elongation of laminated composites is related to the layer thickness size, interface, gradient structure, strain hardening exponent, strain rate parameter and tunnel crack, which are accompanied with crack deflection, crack blunting, crack bridging, stress redistribution, local stress deformation, interfacial delamination crack and so on. The concept of laminated composites can be extended by applying different combination of individual layer, and provides theoretical as well as experimental fundamentals on strengthening and toughening of metal matrix composites.

Keywords: metal matrix composites, multiscale hierarchical structure, laminated structure, rule of average, tunnel crack

1. Introduction

Since the 1980s in the last century, as a result of the significant progress in the material science and technology, many metal matrix composites (MMCs) reinforced with fiber, whisker and particle reinforcements, especially titanium matrix composites (TMCs), have been developed in many applications in order to improve special stiffness, special strength and thermal stability, such as automobile, nuclear, aerospace, armor and sectors and so on [1]. However, MMCs usually exhibit low fracture elongation and toughness in comparison with monolithic metal and alloys due to the addition of reinforcements, which seriously limits their applications [2]. Herein, control of the microstructure and interface characteristics are the important keys to the development of high strength and toughness MMCs and have been achieved by thermomechanical, heat treatment, alloy composition, reinforcement selection and reinforcement surface modification methods [3]. These favorite methods contain many strengthening effects, such as grain refinement, dispersion, solution, work hardening and phase transition strengthening mechanisms. However, grain refinement strengthening is the only method to improve the strength without decreasing fracture elongation, but it is usually no longer suitable when the grain sizes fall below 1 μm [4]. Therefore, it is believed that innovative design of metal matrix composites should make a guide to improve the comprehensive mechanical properties of superior strength, toughness and fracture elongation [2].

Recently, more and more scientists have realized that the nature has provided numerous smart structures for desired mechanical and functional properties after many billion years of stringent evolution [2]. A good idea was inspired from bio-inspired microstructures in nature. Nature materials always reveal multiscale hierarchical structures, such as nacre, which contains laminated, brick, protrusions, bridge and network structures from macroscale, mesoscale, microscale, nanoscale and even the atom-scale as shown in **Figure 1**. They can perfectly delicate themselves at different levels and play a synergistic strengthening and toughening effect during the deformation and fracture process, which results in a superior combination of performances, such as the 3000 times higher toughness of mollusk shells than their individual constituents [5].

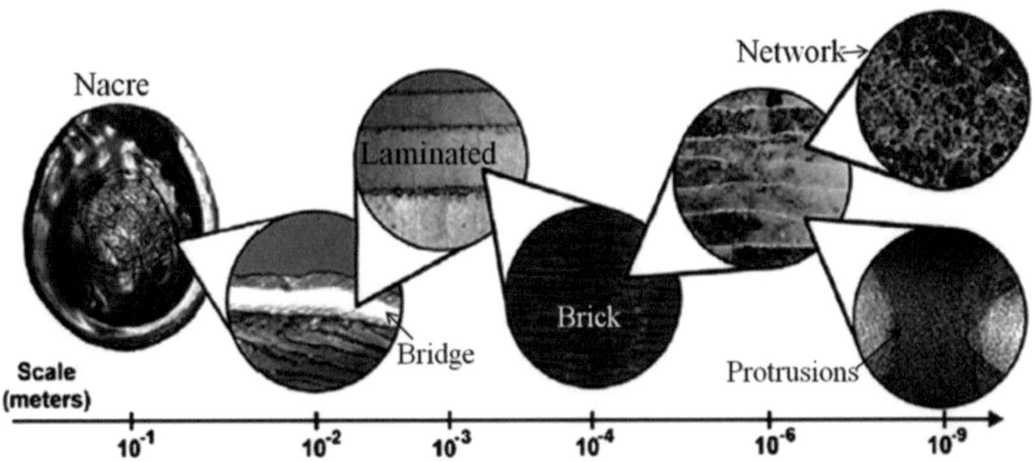

Figure 1. The multiscale hierarchical structures of nacre at several length scales [2, 5].

http://dx.doi.org/10.5772/intechopen.69976

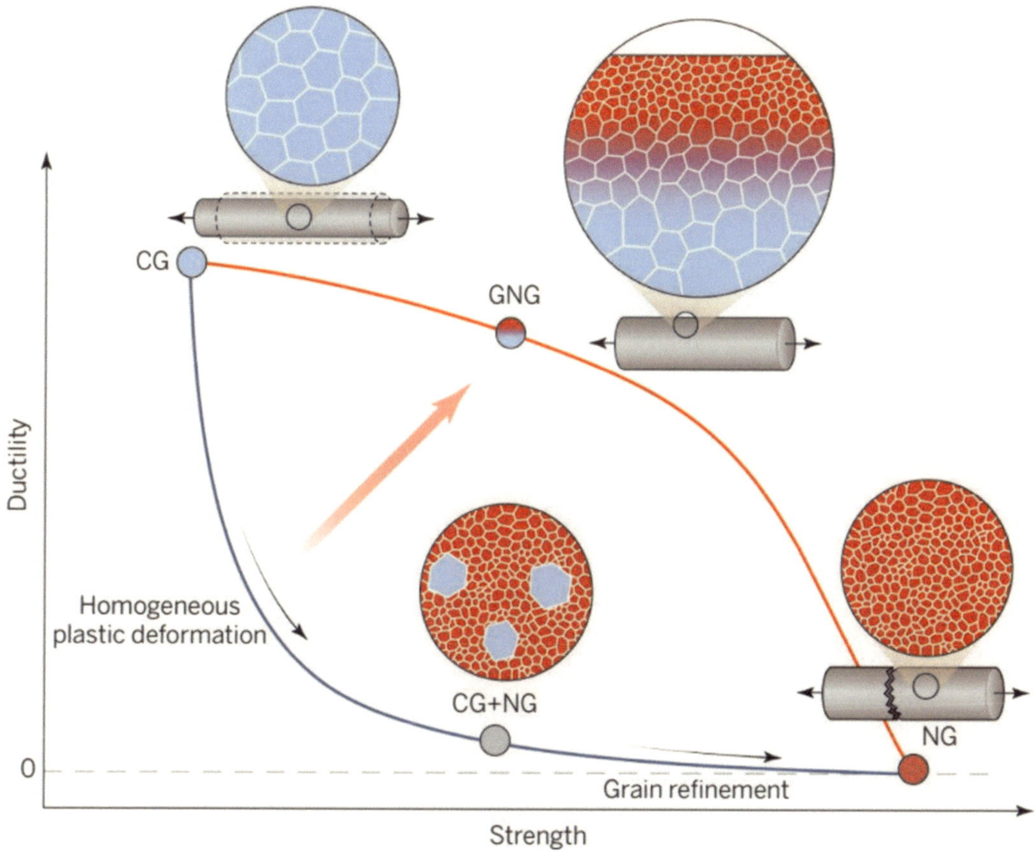

Figure 2. Strength-ductility synergy of metals with gradient grained structure [4].

In 2010, Lu [6] proposed that the metal matrix composites can obtain superior tensile strength and fracture elongation by designing multiscale hierarchical structures. For example, surface mechanical grinding treatment (SMGT) can induce gradient grain distribution and laminated structure in Cu and Ni matrix, respectively. The inspired hierarchical structure is useful to the enhanced fracture toughness of metal matrix composites apart from the monolithic MMCs as shown in **Figure 2** [4, 7, 8]. Palkovic et al. [9] reported that slurry can be designed by the ideal of multiscale hierarchical structures, which can be expected to replace the silicate cement with huge energy consumption. Zheng et al. [10] fabricated multiscale hierarchical metallic metamaterials by three-dimensional (3D) architectures as shown in **Figure 3**, obtaining superior tensile elasticity of 20%.

On the basis of ideal that metal matrix composites with multiscale hierarchical structures can obtain superior mechanical properties, this proposal aims to synthesize laminated Ti-(TiBw/Ti) composites by using powder metallurgy, which contains laminated, network and needle-like structures as shown in **Figure 4**, in order to achieve reinforcing, toughening and plasticizing effect, and to clarify the influence rule of fabrication parameters, laminated, interfacial and network structure parameters on microstructural evolution and mechanical behavior of

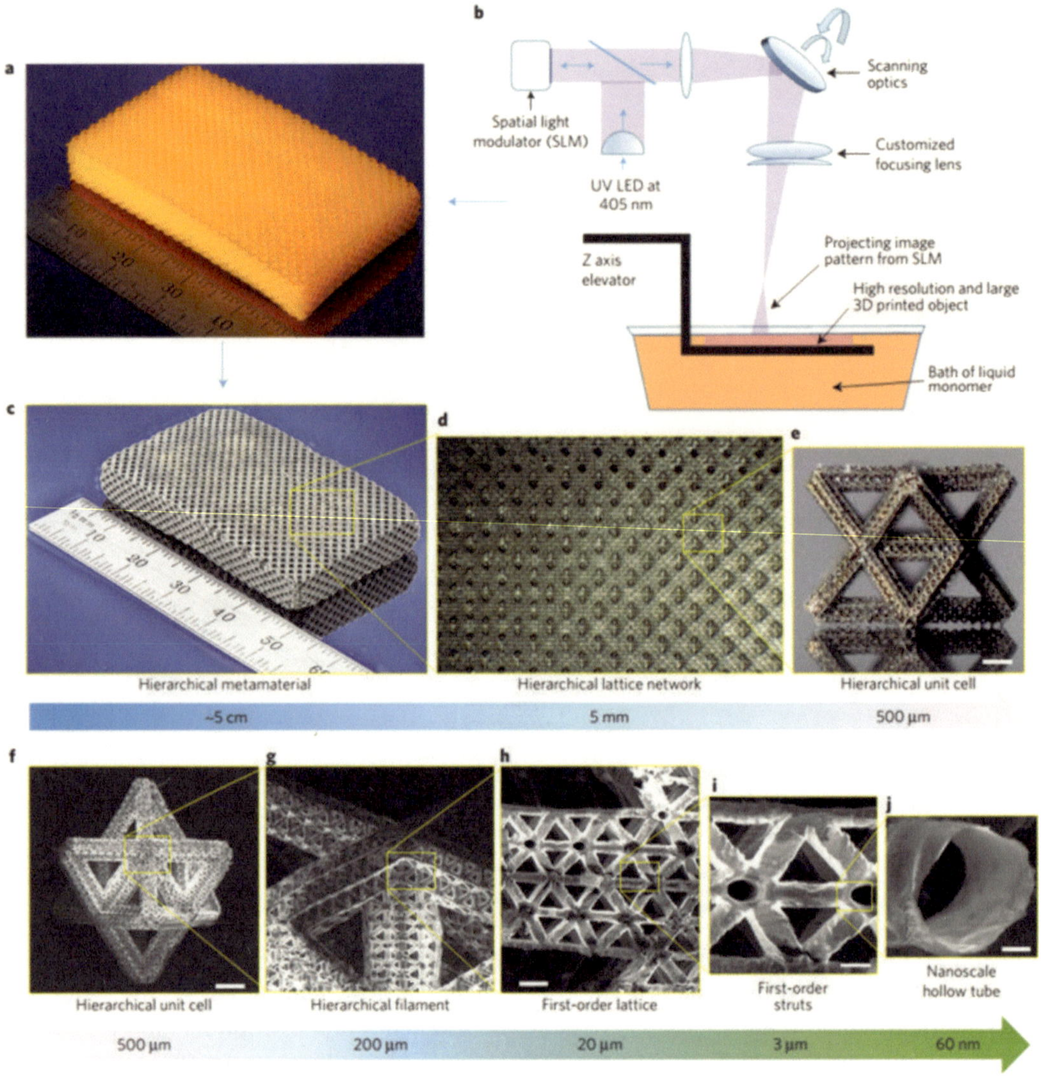

Figure 3. Multiscale hierarchical metallic metamaterials (a-f) [10].

titanium matrix composites. By using tensile and bending tests, the influence regular pattern of sintering temperature, layer thickness, layer thickness ratio and reinforcement volume fraction on mechanical properties should be investigated. It may solve the inverted shortcoming between strength and toughness. Through design and optimization of titanium matrix composites with two-scale hierarchical structures, this work will provide theoretical as well as experimental fundamentals on strengthening and toughening of metal matrix composites.

http://dx.doi.org/10.5772/intechopen.69976

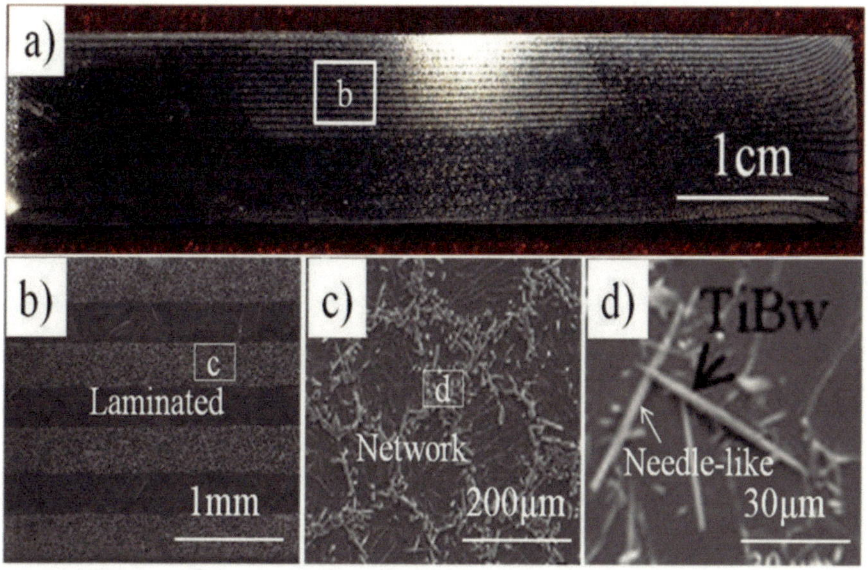

Figure 4. The multiscale hierarchical Ti-(TiBw/Ti) composites. (a) Macrograph, (b) laminated structure, (c) network structure and (d) needle-like structure [2].

2. Multiscale hierarchical structure

Yin [11] concluded that the one-scale hierarchical distribution structure of reinforcements contains dual matrix, network, laminated, gradient structural and other bi-continuous types. **Figure 5** shows the four structures in the titanium matrix composites reinforced with TiBw. Morsi and Patel et al. [12–14] fabricated a so-called "dual matrix" Ti-TiB composites as shown in **Figure 5(a)**, which contains isolated reinforcement-rich phase and continuous Ti matrix phase, exhibiting superior fracture toughness. Huang et al. [2, 15, 16] fabricated one-scale network hierarchical TiBw/Ti composites and two-scale network hierarchical TiBw + Ti_5Si_3/Ti composites as shown in **Figure 5(b)**. The elastic modulus, passion ratio and yield strength reach to the upper limit of Hashin-Shtrikman theory, as well as high oxidation resistance, superior fracture elongation, toughness and tensile strength. Han et al. [17] fabricated the laminated Ti-(TiB + La_2O_3)/Ti composites by powder metallurgy and hot rolling process as shown in **Figure 5(c)**, which results in an improvement of 30% in elongation compared with monolithic composites with same volume fraction of reinforcements. In 2003, Panda et al. [18–21] fabricated TiB/Ti and TiB_2/Ti functional gradient composites by powder metallurgy as shown in **Figure 5(d)**, which effectively weakens the thermal residual stress and inhibited the formation of cracks.

Figure 6 shows the fracture characteristics of the "dual matrix" Al_2O_3/Cu composites. The crack propagated along the Cu matrix and Al_2O_3 ceramic interface or inside the Cu matrix as

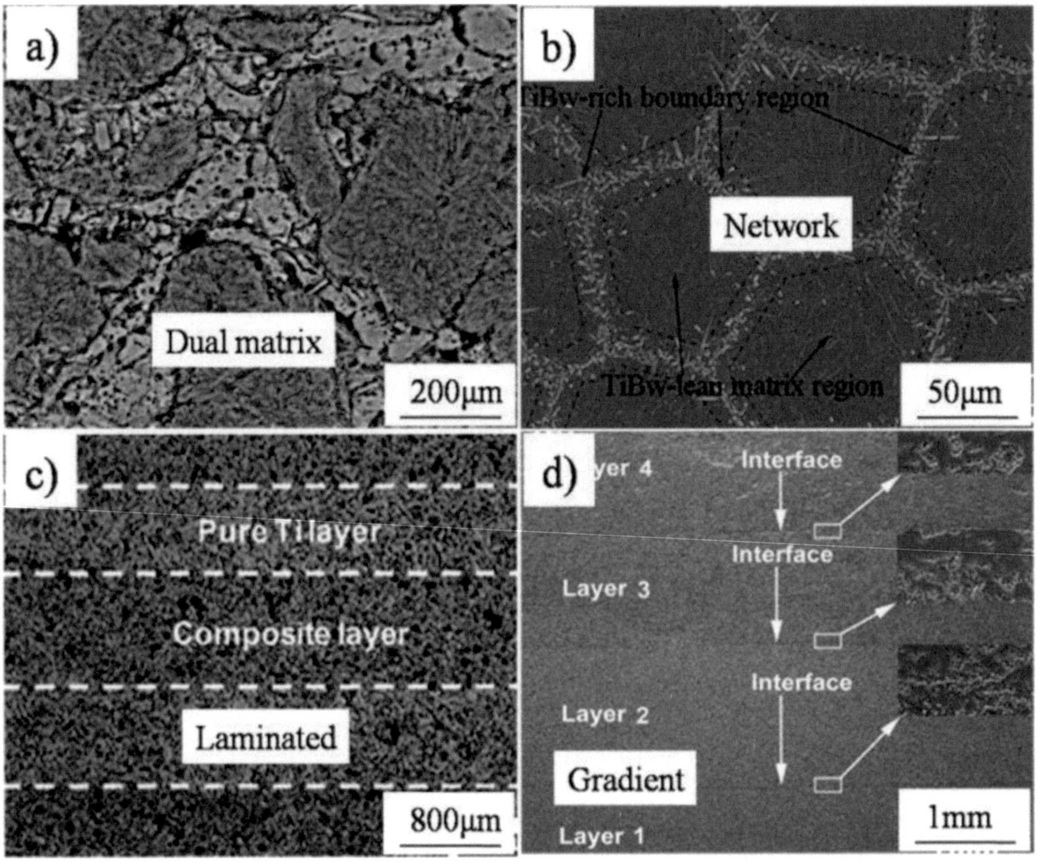

Figure 5. Four structural types of TiBw/Ti composites. (a) Dual matrix [12], (b) network [2], (c) laminated [17] and (d) gradient structure [18].

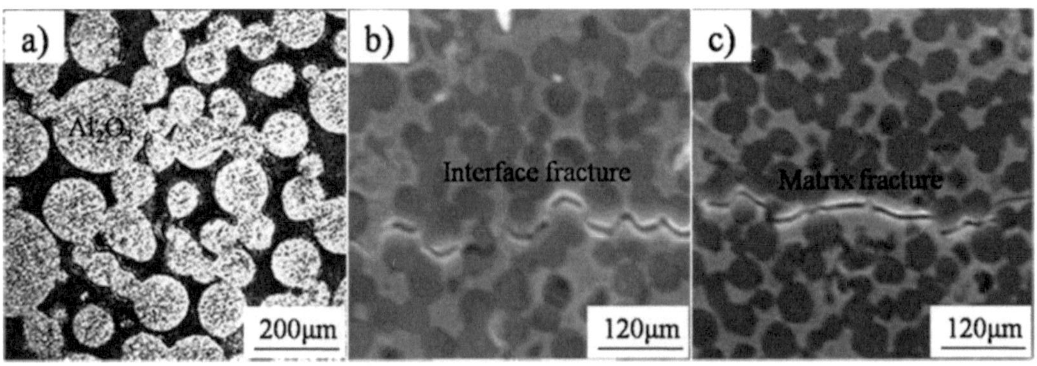

Figure 6. The fracture morphology of Al_2O_3/Cu composites [22]. (a) Microstructure, (b) interface fracture and (c) matrix fracture.

shown in **Figure 6(b, c)**. The two tortuous propagation paths absorbed a lot of fracture energy and plastic deformation work, respectively, which are benefit to the high fracture toughness compared to the popular belief that crack propagation occurs in the brittle ceramic phase. Herein, the interface fracture is attributed to the high thermal tensile stress at the interface developed during the fabrication process [22].

Huang et al. [2, 23, 24] concluded that the titanium matrix composites with network structure correspond to the upper limit of Hashin-Shtrikman (H-S) bound. Based on H-S bounds proposed by Hashin and Shtrikman in 1963 [25, 26], the reinforcement with network structure can be viewed as "stiffer" phase, which is continuous and encapsulates the titanium matrix as shell structure. The elastic modulus of this structure can be predicted by the H-S upper bounds. However, the titanium matrix composites with homogeneous distribution of reinforcement can be taken as that the soft phase encapsulates the stiffer phase, whose elastic modulus is near to the lower limit of H-S bound, and the H-S bounds can be expressed as follows [2]:

$$E_{HS-upper} = \frac{E_w[E_w V_w + E_m(2 - V_w)]}{E_m V_w + E_w(2 - V_w)} \quad (1)$$

$$E_{HS-\text{lower}} = \frac{E_m[E_m(1 - V_w) + E_w(1 + V_w)]}{E_w(1 - V_w) + E_m(2 - V_w)} \quad (2)$$

where E_w and E_m are the elastic modulus of whisker reinforcement and matrix, respectively, V_w and V_m are the volume fractions of the whisker and matrix, respectively. The network structure can enhance the strengthening and toughening effect at room temperature, which is similar with the grain-boundary strengthening mechanism. At high temperature, it can also overcome the grain-boundary weakening behavior. Therefore, TiBw/Ti composites with network structure obtain superior room and high-temperature properties as shown in **Figure 7**.

Figure 8 shows the illustration diagram of oxidation and fracture resistant mechanisms of titanium matrix composites with network structure [2]. After oxidation process of 140 h at 973K, the surface of titanium matrix composites with network structure is still smooth without any spallation and cracks. The in situ reacted TiC particle forms a dense continuous wall as shown in **Figure 8(b)**, which can effectively inhibit the oxidation velocity as shown in **Figure 8(c)** [2, 27].

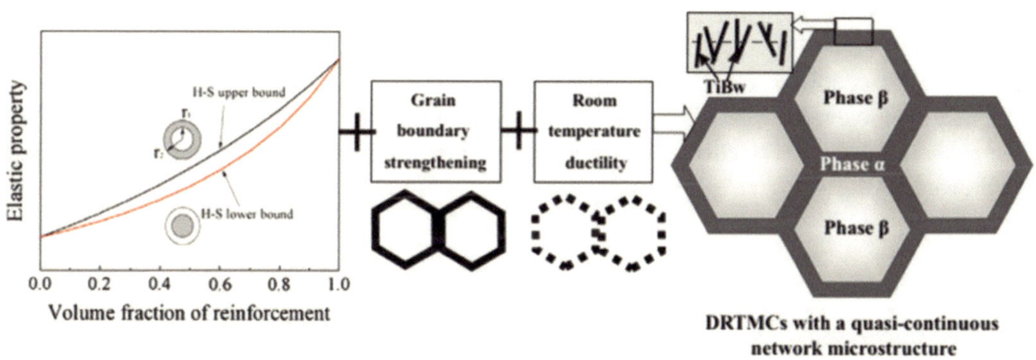

Figure 7. It shows the strengthening and toughening mechanisms of TiBw/Ti composites.

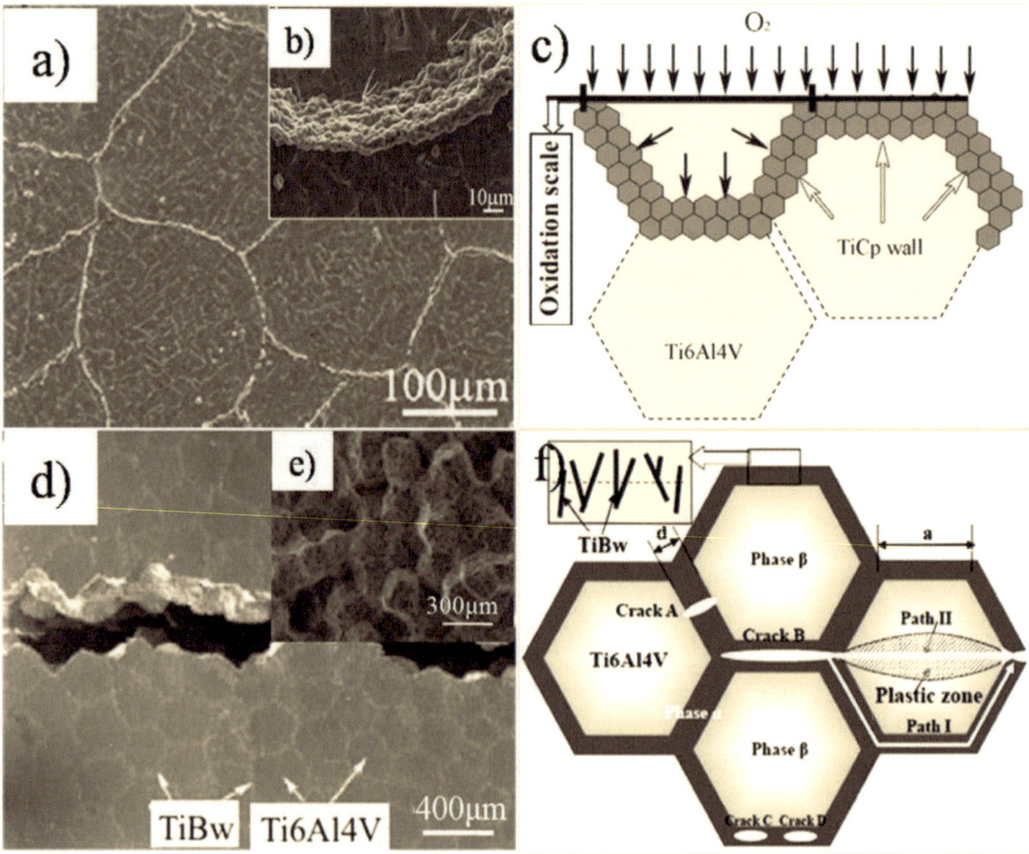

Figure 8. Schematic illustration showing the oxidation and fracture resistant mechanisms of titanium matrix composites with network structure [2]. (a–c) Oxidation and (d–f) fracture resistance.

During the fracture process, the TiBw/Ti6Al4V composites are mainly fractured at the network boundary as shown in **Figure 8(d, e)**, and the matrix tearing is also the fracture characteristics with low volume fraction of reinforcement. **Figure 8(f)** illustrates the crack propagation paths and toughening behaviors of TiBw/Ti6Al4V composites with low volume fraction [2]. Crack A shows that the crack perpendicular to the network boundary cannot propagate through the softer matrix phase because of low crack size, which can play strengthening and toughening effect in the composites. However, the different propagation paths along path I or path II mainly depend on their absorbing energy. The effect reinforcement strength and absorbing energy can be expressed as follows [2, 28]:

$$\sigma = \sqrt{\frac{\pi EG}{2(1-\nu^2)d}} \tag{3}$$

$$Q = \sigma.S.L = AL\sqrt{\frac{\pi EG}{2(1-\nu^2)d}}.d^2 = AL\sqrt{\frac{\pi EG}{2(1-\nu^2)}}.d^{\frac{3}{2}} \tag{4}$$

where σ is the effective reinforcement strength and G is the critical strain energy release rate for the dynamic propagation of a crack. ν is the Poisson's ratio, d is the reinforcement size along the crack direction, S is the crack surface area, L is the open crack length and A is the parameter related to the surface and length.

The absorbing energy of crack propagation along path I (Q_I) and II (Q_{II}) can be deduced as Eqs. (5) and (6), respectively [2].

$$Q_I = AL \cdot B\sqrt{\frac{\pi E_I G_I}{2(1-\nu_I^2)}} \cdot (3a)^{\frac{3}{2}} \tag{5}$$

$$Q_{II} = AL \cdot B\sqrt{\frac{\pi E_{II} G_{II}}{2(1-\nu_{II}^2)}} \cdot (2a)^{\frac{3}{2}} \tag{6}$$

Therefore, the crack can propagate along path I when $Q_I \leq Q_{II}$.

The concept of functionally gradient Ti-TiB composites arose as a result of efforts to prevent the cracking in TiB-Ti joints induced by the coefficient of thermal expansion (CTE) mismatch during high-temperature processing. Functionally gradient materials (FGM) effectively decrease the

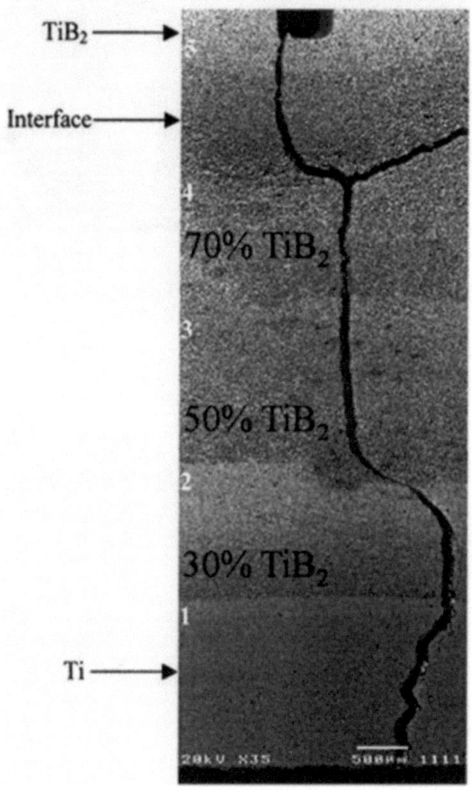

Figure 9. The crack propagation of functionally gradient TiB-Ti composites [29].

thermal stresses and delay the plastic yielding and the fracture failure of the metallic layer by the load shearing of the ceramic [18–21]. Meanwhile, it is a better overall use of the available materials by distributing them in a suitable spatial direction. Ma et al. concluded that the main crack deflected to the layer interface, and it is interesting to note that crack bifurcation was initiated at the interface between layers 5 and 4, resulting into the stepwise fracture process rather than catastrophic failure as shown in **Figure 9**. Moreover, it has good ballistic properties that make it popular armor application [21, 29].

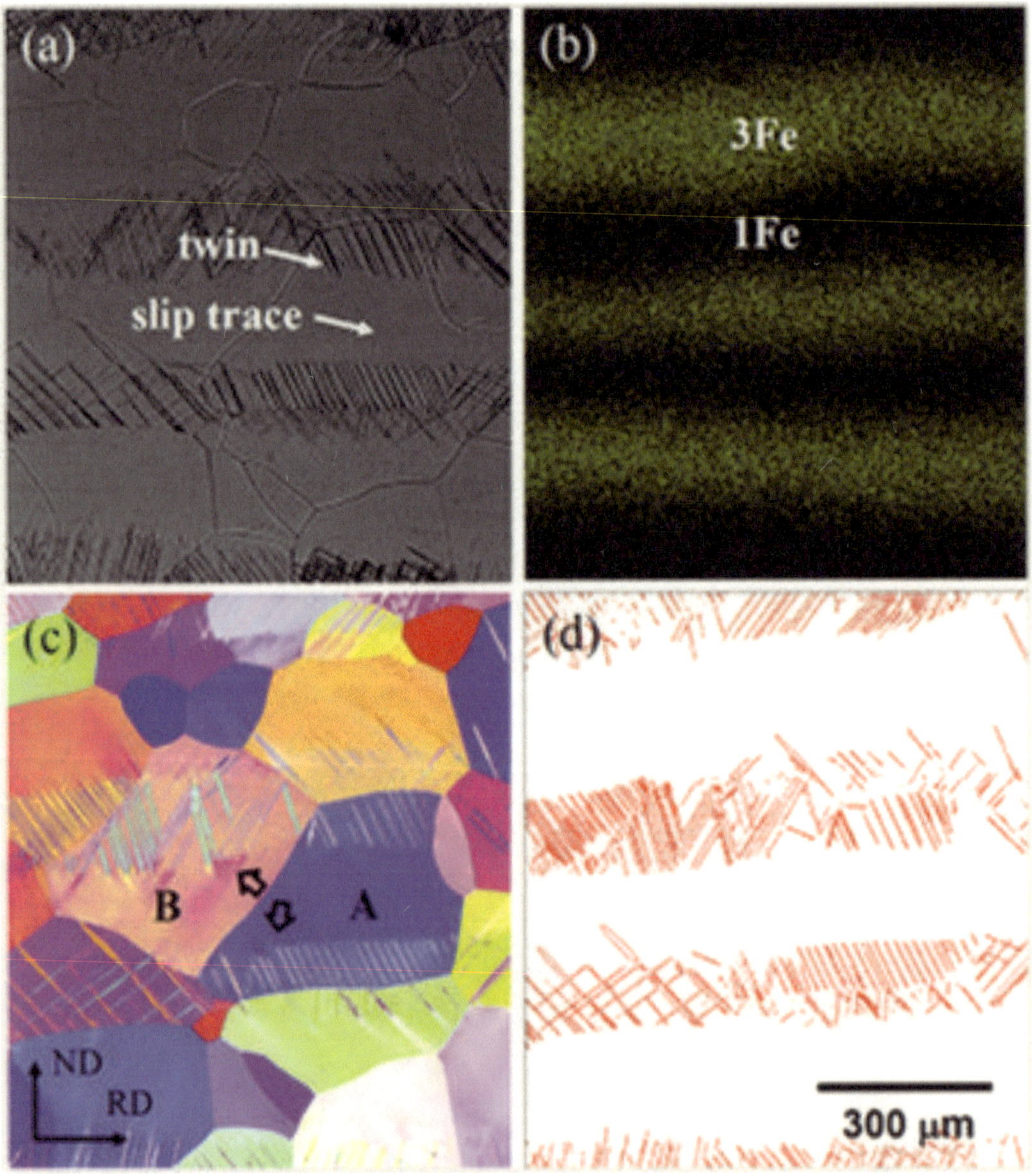

Figure 10. (a–d) {332}<113> Twinning and dislocation deformation of laminated Ti composites [30].

Min et al. [30, 31] fabricated a multilayered Ti-10Mo-1Fe/Ti-10Mo-3Fe composites by hot roll bonding and subsequent heat treatment as shown in **Figure 10(a)**. The layer interface can be detected by Fe concentration as shown in **Figure 10(b)**. It is interesting to note that Ti-10Mo-1Fe layer was deformed by twinning, whereas deformation characteristic of Ti-10Mo-3Fe is dislocation slip as shown in **Figure 10(c, d)**. The artificially multilayered twinning/dislocation slip deformation characteristics can provide a high strength-ductility combination over a wider range to meet industrially applicable strength and formability requirements.

3. Laminated structure

Actually, composite materials with laminated structures have been a design ideal to improve the mechanical and functional properties from antiquity. "The Iliad" of homer has described that Achilles's shield has five layers, and the sequence of laminates is bronze/tin/gold/tin/bronze. The Aeneas's spear penetrated the first two layers but trapped by gold layer during combat, revealing superior mechanical properties. In 1837, an iron-laminated composite dating as far back as 2750 BC was found in the Great Pyramid in Gizeh, Egypt, and many layers of wrought iron have been inexpertly welded tougher by hammering as shown in **Figure 11** [32, 33]. Lamination architect ideal may be attributed to the following three reasons: (1) only thin sheets could be carburized in the earliest form of wrought iron, so lamination was a good way to create bulk composite materials; (2) it is no expensive to sandwich hard rare steel between more cheap common materials and (3) optimizing the combination of strength, toughness and sharpness is the basis for lamination from the mechanical view of materials. Therefore, many weapons were made by lamination design from ancient, such as Adze blade, Japanese sword, Merovingian blade and Chinese blade [33].

Since 1988, American scientists took the sandwich materials into the aircraft door C-17, and the investigations of laminated and functional gradient metal composites enter into the period of rapid development [34]. Jiang et al. [35] fabricated laminated Ti-Al_3Ti composites using reaction vacuum sintering in 2003. During the impact testing process, the laminated composites can absorb lots of impact works through forming delamination cracks, and it is expected to be applied in the armor equipment and ballistic system. In 2006, Japan government sponsored a research project named "the layer integrated steel and metals," which aims to fabricate multi-layer steel with high strength (>1200 MPa) and superior tensile elongation (>20%) through the composite design of high strength steels and ductile steel layers [36]. In 2008, Inoue et al. [37] fabricated multilayered SS420/SS304 steel with high toughness and reported that fracture elongation is gradually increasing with the decrease of layer thickness, which is attributed to the decreasing size of tunnel cracks. Nambu et al. [38] reported that multilayered steel with strong interfacial bonding strength exhibits superior longitudinal fracture elongation compared with the laminated composites with weak interfaces. Moreover, they found that the laminated structures can change the deformation mode of martensite layer, transiting from poor work hardening to improved hardening [39].

Strength and toughness are the most important performance in traditional structural materials [1–3, 40]. Recently, with the development of national defense and civil sectors, such as aerospace, nuclear, shipbuilding, automobile and so on, a high demand of strength and

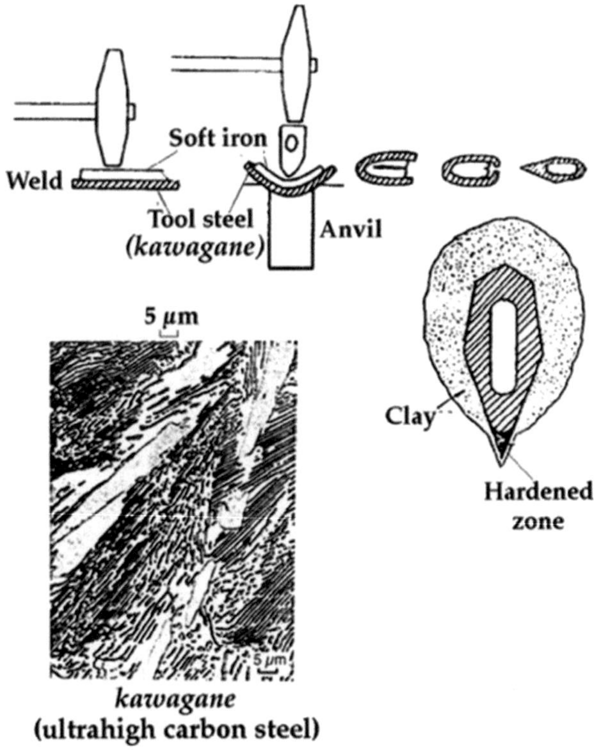

Figure 11. The fabrication process of slip blacksmiths [32, 33].

toughness is necessary for structural materials to adapt the increasingly stringent environment and application conditions [4, 6, 41]. For centuries, scientists have been conducting investigations to develop composite materials with laminated structures to improve strength and toughness [32–34, 42]. Typical laminated composites, such as ceramic-ceramic [43], metal-metallic glass [44], metal-ceramic [45], metal-metal [46], metal-metal composites [47] and metal-ceramic-intermetallic [48] systems have shown desirable structural properties as a result of many layers and interfaces, and these laminated composites allow for the possibility of combining the high ductility and toughness of soft layers with the high strength of hard layers.

Though the design concept, strengthening and toughening mechanisms of laminated composites are ancient and attempted for a long time, the relationship between microstructure and mechanical properties are still unclear and ambiguous. Lesuer et al. [32] concluded that the elastic property and yield strength of laminated composites follow the "rule of average." However, the estimation of fracture elongation and fracture strength is complex, which is inconsistent with the "rule of average" as shown in **Figure 12** [40]. Therefore, the further investigation of the fabrication and structure factors of laminated composites may prove useful to understand the relationship between structure and performation. The results could be used as guide to design microstructures and improve mechanical properties by multilayer structure, which require a completely different approach from the investigations of conventional monolithic composite material.

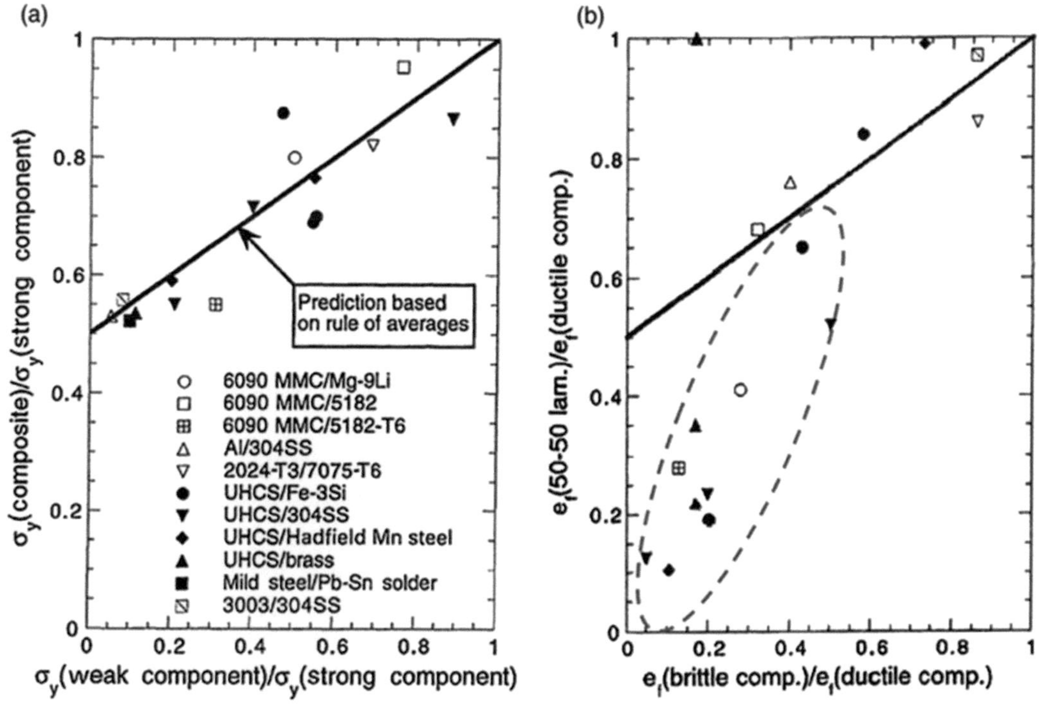

Figure 12. (a) Yield strength and (b) elongation of different metals consisting of 50% strong (of brittle) component and 50% weak (or ductile) component [40].

4. Strengthening and toughening mechanisms of laminated structure

Typical laminated composites have shown desirable mechanical properties, such as tensile strength, fracture toughness, compact toughness, formability, bending properties and fracture elongation. A number of papers have been published on the reinforcing and toughening of laminated composites and illustrated that laminated metal composites are especially effective in improving fracture-related properties by various intrinsic and extrinsic mechanisms [32, 49].

Recently, the strengthening and toughening mechanisms of laminated composites mainly focus on the investigation of plastic deformation, crack propagation and fracture morphology. Herein, Cetin and Hutchinson et al. [50–52] carried out the depth investigation on the tunnel cracks, plastic bifurcations and delamination cracks and found that the fracture elongation is related to the layer thickness, layer thickness ratio, yield strength, ultimate strength, interfacial bonding strength, work hardening index, strain rate sensitivity exponent and interfacial reaction phase of individual layer. Serror [53] reported that multilocalized necking of laminated composites is benefit to relieve the localized stress concentration and improve plastic deformation capability during maximum tensile testing process based on the plastic bifurcation theory and finite element method (FEM) as shown in **Figure 13**. Zhang et al. [54] fabricated multilayer Cu/Al sheets using rolling bonding method and found that laminated structure can effectively delay the premature localized necking behavior of Cu sheet. They also investigate that nanometer indentation behavior and found that size effect on the Cu/Cr, Cu/Au and Cu/Ni multilayered composites by magnetron sputtering. The ultimate strength would obviously be enhanced with the decreasing layer thickness in the nanoscale. However, the ductility and fracture toughness of laminated composites are ambiguous due to the complex interaction between layer thickness and interfacial structure [55]. The contribution of this section is to draw the series of deformation and fracture mechanisms of laminated composites and to provide guide for the reinforcing and toughening effect research and design.

4.1. High fracture toughness of laminated composites

The typical natural laminated composite is nacre material with the block-wall structures as shown in **Figure 1**. The nacre contains mineral layer ($CaCO_3$) with the volume fraction of 99% and protein with the volume fraction of 1%. Nacre reveals high toughening mechanisms as follows: (1) the layer boundary with many projections reveals obvious bridging reinforcing effect; (2) interfacial delamination can play toughening effect when loaded on the nacre; (3) the staggered block-wall microstructures with a high sheet lap length and length/diameter ratio can decrease the relative displacement of fiber sheets during fracture process; (4) the mineral layer can be bonded by protein with three-dimensional network structure and (5) the nacre with convex-curved macroscopic morphology can state biaxial stress state, and the convex-curved plane only affords average compression stress, seldom of tensile stress and bending stress, which is a benefit to the ceramic matrix composites [5].

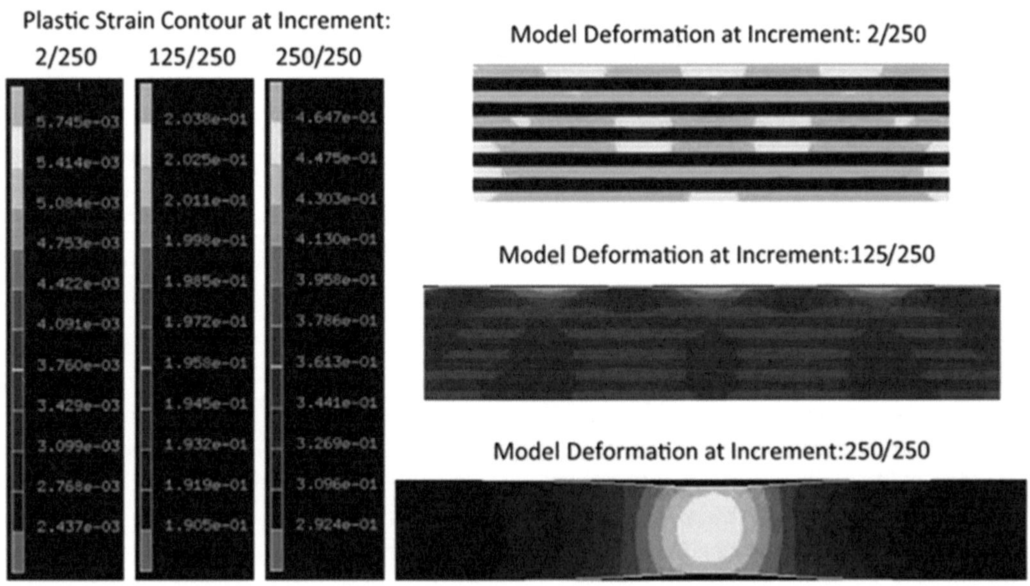

Figure 13. Three snapshots of the deformation and corresponding plastic strain contours of laminated composites under plane strain uniaxial tension [53].

Figure 14 shows the mechanical properties and toughening mechanisms of laminated ceramics with different interfacial status and constitute layers [56, 57]. Two kinds of laminated ceramic composites, the laminated composites with strong interface and weak interface, can be divided according to interfacial bonding strength. When the laminated ceramic composites afford bending force, the two kinds of laminated composites always improve the bending properties and fracture toughness through different toughening types. The laminated ceramic composites with strong interfacial strength possess high residual stress during fabrication process, and the main crack may appear to deflect or bifurcation on the strong interface during the fracture process induced by residual compression stress. Therefore, crack propagation path can be extended and the stress concentration at the crack tip can be mitigated, so the laminated ceramic composites with strong interfacial bonding strength can obtain high fracture toughness. The laminated ceramic composites with weak interfacial strength can also obtain high fracture toughness by delamination crack. When the main crack propagates into the weak interfaces forward, it also deflects or forms interfacial crack along the weak interface due to large amount of defects. The stress concentration can be relieved and the stress state can be changed into biaxial stress from complex triaxial stress, which resulted into the shielding phenomenon of main crack tip [58].

The improvement in fracture toughness of laminated ceramic composites can be mainly attributed to two categories: intrinsic and extrinsic mechanisms as shown in **Figure 15** [59, 60]. Intrinsic toughening mechanism implies in situ resistance of the microstructure (precipitates, particle spacing, grain size, morphology and so on) on crack propagation. For example, toughness can be enhanced by changing the nature, distribution and/or interface properties of second phase particles to suppress damage in the form of microcracking or microvoid

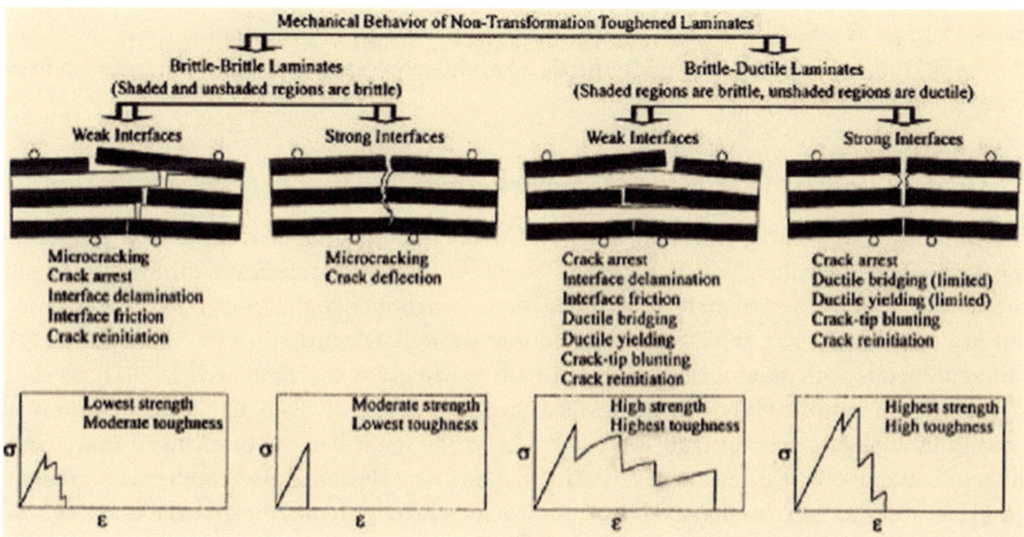

Figure 14. The mechanical properties and toughening mechanisms of laminated ceramics with different interfacial status and constitute layers [57].

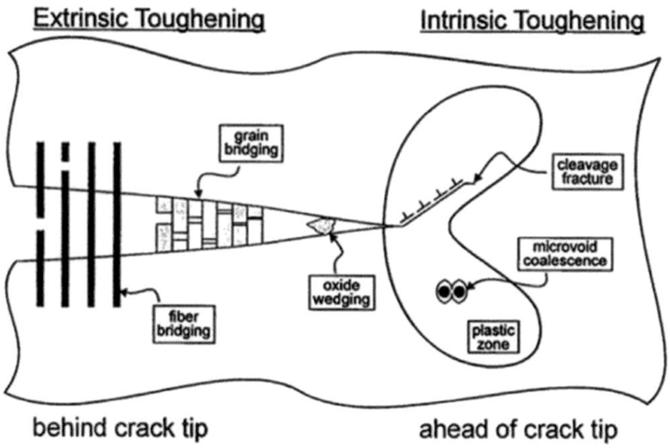

Figure 15. The intrinsic and extrinsic toughening mechanisms [60].

formation ahead of the crack tip. Extrinsic toughening mechanism indicates the microstructure behind of the crack tip to effectively decrease the crack driving force, which is initiated by the residual stress that reduces local stress intensity at the crack tip, such as delamination at weak interfaces [45], crack bridging [61], stress redistribution [32, 33], crack front convolution [32], local front convolution [33], crack trapping [48], ductile ligament [60], buckling [62], multiple cracking [63], crack bifurcation in the compressive stress layer [60], phase transformation under stress and compositional gradient [59] and crack deflection by mismatch of thermal expansion coefficients between different layers inevitably generates thermal residual stresses during subsequent cooling of laminated composites [64]. These behaviors need a significant amount of energy absorption and long crack propagation paths, resulting in an increase in fracture toughness as shown in **Table 1** [33]. Therefore, extrinsic mechanisms, on the other hand, improve fracture properties by providing alternative crack propagation routes or reducing the driving force for crack growth through various processes that shield the crack tip from the applied stress [65].

4.2. Gradient laminated distribution of microstructure

It is all known that metal with nanoscale grains can obtain superior compression properties but low tensile ductility, which is due to premature strain localization during the tensile processing [4, 6]. However, metal with coarse grains can obtain high tensile fracture elongation but low tensile strength, which is attributed low strain hardening capacity. What happens if different metals with nanometer grains and with coarse grain are combined? In 2011, Lu et al. [7] introduced nanosize layer on the surface of copper during the serious plastic deformation process by surface mechanical attrition technology. The grain size can be changed from nanoscale into microscale with the increase of depth gradually. The laminated copper with gradient grain size distribution can obtain superior tensile yield strength and high tensile ductility, and the tensile ductility can reach to 100% without crack initiation. In 2014, Liu et al. introduce high-rate shear deformation with strain gradients on bulk nickel, forming nanometer-scale laminated structure with 20 nm, which exhibits a superior hardness of 6.4 GPA [8]. Lu

http://dx.doi.org/10.5772/intechopen.69976

Mechanism (Testing orientation)	Volume fraction dependence	R-curve behavior possible	
Crack deflection (Crack arrester)	No	. . .	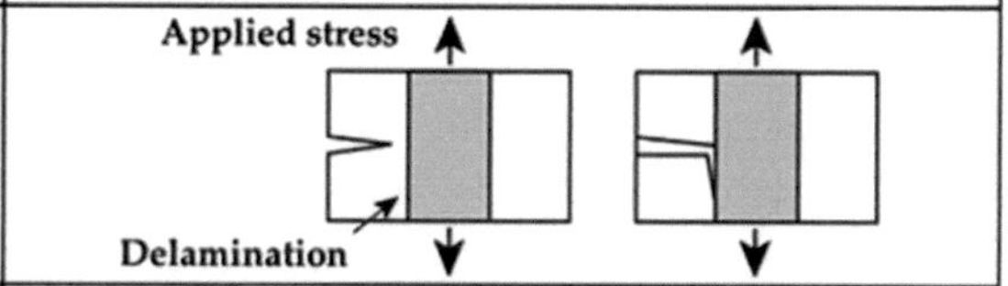
Crack blunting (Crack arrester)	No	. . .	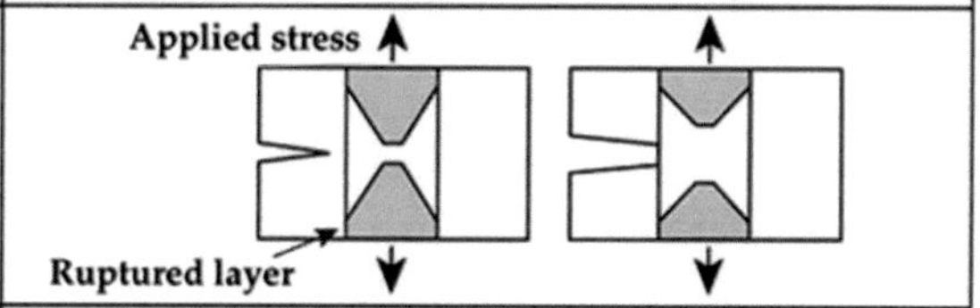
Crack bridging (Crack arrester)	Yes	Yes	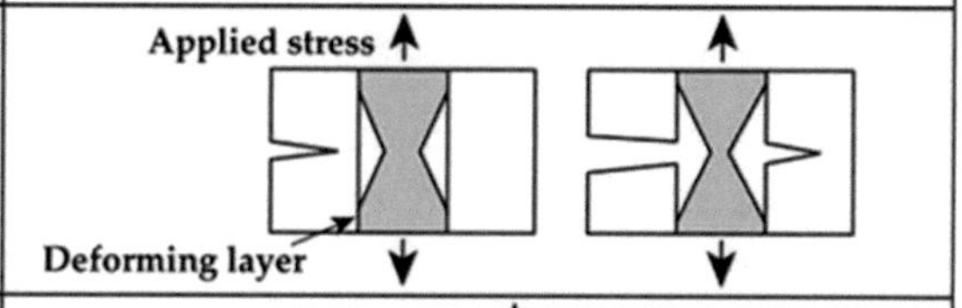
Stress redistribution (Crack arrester, crack divider)	. . .	Yes	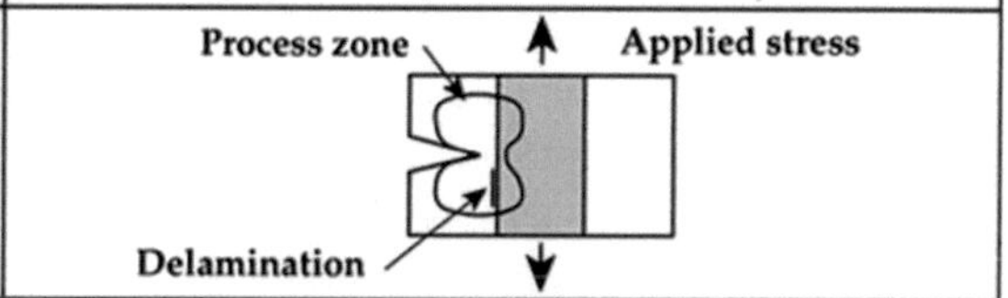
Crack front convolution (Crack divider)	Yes	Yes	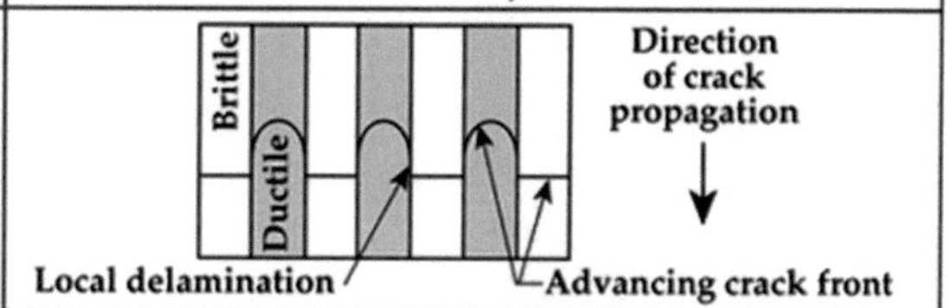
Local plane stress deformation (Crack divider)	. . .	Yes	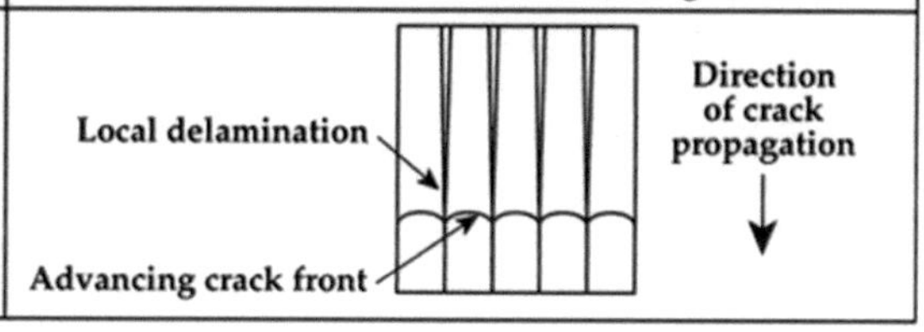

Table 1. Toughening mechanisms for laminated metal composites [33].

proposed that the gradient nanometer grained surface layer results in synergy effect of strength and ductility rather than the tradition trade-off between strength and ductility [4].

4.3. Size effect in the laminated composites

Gao et al. [66] reveal that the laminated composites become insensitive to flaw when the layer thickness is lower than the critical flaw size according to the research on nacre. The stress field becomes more and more uniform as the layer thickness decreases. The critical flaw size is expressed as follows:

$$L_{critical} = \frac{\alpha \Gamma E}{\sigma^2} \tag{7}$$

where α is crack propagation parameter, about 1.3–1.9. E is the Young's modulus, and Γ is the fracture energy of atomically sharp cracks. Kumar et al. proved that when the layer thickness is small enough compared to the critical flaw size, $L_{critical}$, no matter how high value of applied stress, even reach the ultimate strength of the material and the laminated composites will rupture wholly rather than classical damage mode, which experience nucleation, coalescence, growth and step-by-step crack propagation path. Once layer thickness is beyond the $L_{critical}$ and tunnel crack forms, the laminated composites will damage like domino, which does not need a lot of fracture energy. However, the wholly fracture process will spend more fracture energy. Therefore, laminated composites with small layer thickness can obtain more fracture strength and fracture toughness.

However, most experimental results reveal that the ductility of laminated composites is seriously limited with reducing layer thickness to a critical value (<100 nm), although fracture strength has been shifted. Zhang et al. [55] indicated that the fracture mode of the laminated Cu/Au and Cu/Cr composites changed from ductile shearing to brittle shearing with decreasing layer thickness as shown in **Figure 16**. Laminated composites with thick layer thickness exhibit high plastic deformation behavior, which is attributed to dislocation multiplication and gliding mechanism, until stress concentration at the interface due to the dislocation piling up, and the yield strength is consistent with the Hall-Patch equation in microscale. However, the interfacial bonding strength gradually decreases with decreasing layer thickness, and the dislocation can transmit across the layer interface to accommodate plastic deformation when the shear stress exceeds the interfacial bonding strength; the yield strength can be estimated by Hall-Patch equation. Therefore, the transition from ductile fracture to brittle fracture may be attributed to the layer thickness size, interfacial constraining effect and columnar grain orientation.

According to Hsia et al. [67] theoretical model as shown in **Figure 17**, the plastic deformation zone size (r_p) at the onset of crack tip is expressed as follows:

$$r_p = \frac{1}{2\pi}\left(\frac{K_{Ic}}{\sigma_s}\right)^2 \tag{8}$$

where K_{Ic} is the fracture toughness and σ_s is the yield strength of the metal. Therefore, there is no confinement exists, and the fracture energy is independent of layer thickness when the layer thickness (h) is higher than $2r_p$. However, the plastic deformation behavior becomes confined when $h \le 2r_p$, resulting in a lower fracture energy. Meanwhile, the tensile stress at the crack tip increases as h decreases. When h approaches the micro level, the dislocation activities within the thin ductile layer can no longer be treated with continuum plasticity theory, and the confined ductile layer with thin thickness makes dislocations pile up at interface, and even the preexisting dislocations with low number are unlikely to blunt the tunnel crack tip, which results in the transition from ductile to brittle behavior

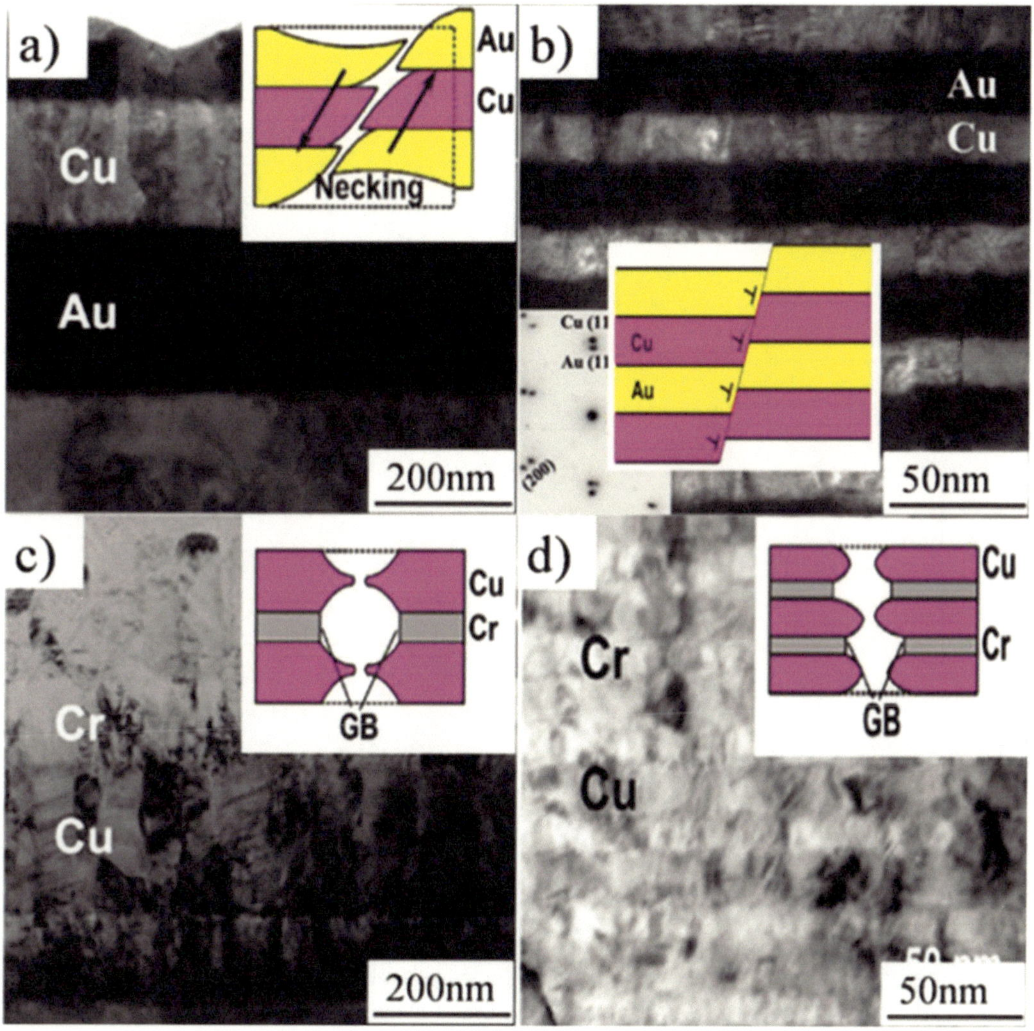

Figure 16. Size effect of laminated composites with layer thickness of (a, b) 200nm and (c, d) 50nm [55].

with decreasing ductile layer thickness by introducing the laminated structure, the confined layer thickness makes dislocations pile up at interface. A low equilibrium number of dislocations can be obtained in the thin ductile layer thickness, which leads to the super-high tensile stress and low fracture toughness at the crack tip, and even cleavage occurs when the tensile stress reaches the theoretical fracture strength. An equilibrium number of dislocations exists at a given load level and ductile layer thickness. Dislocations emitted from the crack tip can not only blunt the crack, but also sent a back stress to the crack tip. Above two effects are mutually competing each other; the former reduces the tensile stress at the crack tip, whereas the latter prevents more dislocation emissions and results into the increase of tensile stress.

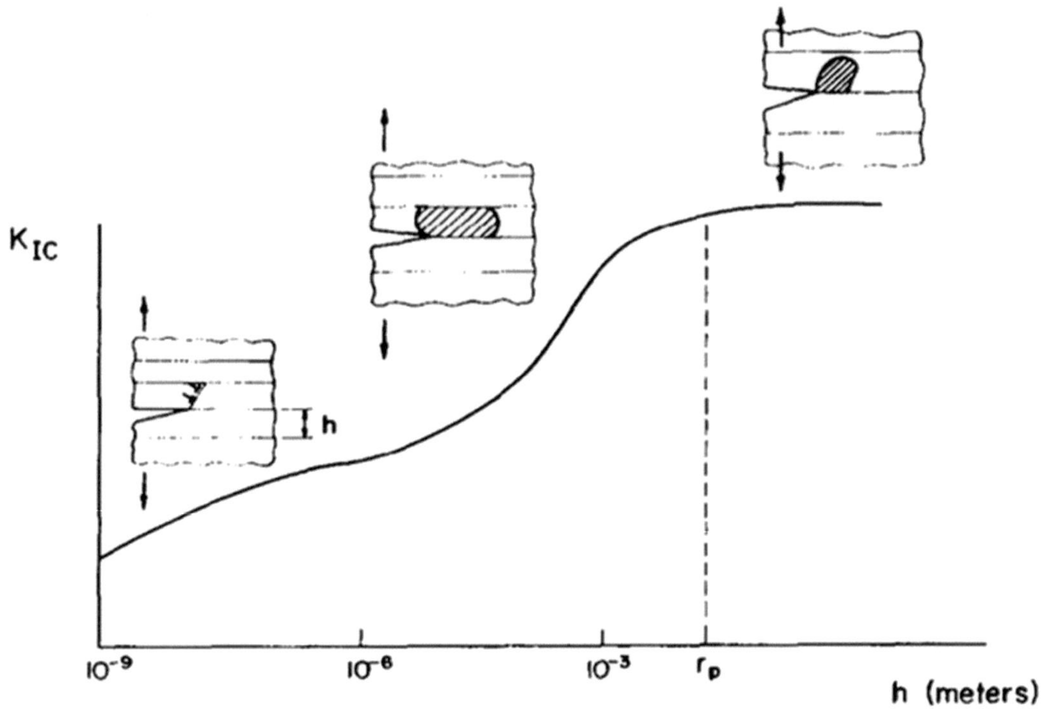

Figure 17. Schematic illustration of fracture energy as a function of layer thickness [67].

4.4. Interface layer

4.4.1. Interfacial bonding strength

Nambu et al. [38] reported that laminated composites with strong interfacial bonding strength exhibit superior longitudinal fracture elongation compared with the laminated composites with weak interfaces. For a metal film deposited on an elastic substrate, Li et al. [68] predicted that strong interface can delay strain localization by finite element simulation (FEM). However, if the metal film approaches to form interfacial delamination crack due to weak interfacial bonding strength, the metal film becomes freestanding and is easy to form a localized necking. Koseki et al. [40] proved that delamination crack is related to the low interfacial toughness, while laminated composites with high interfacial toughness reveal high fracture elongation before failure, leading to localized necking and ductile fracture surface, exhibiting interfacial delamination crack and brittle fracture of brittle layer when the interface becomes unstable with the increase of strain. The debonding criterion is expressed as follows:

$$\Gamma_{int} \geq 0.26 \frac{\pi t_B \sigma^2}{2E_B} \quad (9)$$

where Γ_{int} is the interfacial toughness and σ is the stress imposed on the brittle layer. This criterion provides a reasonable conditions of sufficient interfacial toughness to inhibit the interfacial delamination phenomenon.

4.4.2. Interfacial reaction phase

The interfacial reaction between two different layers can result in decreasing mechanical properties of laminated composites [50]. There are a series of studies on the mechanical response of laminated composites based on Ti-Al [69], Ti-Cu [70], Ti-Ni [71], Ni-Al [72], Cu-Al [73], Ti-Nb [74], Fe-Al [75, 76] and Fe-Cu [77] systems. **Table 2** shows the room temperature tensile properties of these laminated composites. The laminated composites reveal low fracture elongation compared to the constituent metals, which is attributed to the sintering voids and in situ reacted brittle intermetallic layer.

The size of reacted intermetallic layer can also affect the mechanical properties of laminated composites. Pang et al. [78] and Huang et al. [79] reveal that laminated Ti-(SiCp/Al) and Ti/Al composites exhibit high fracture ductility due to the in situ reacted nanosized $TiAl_3$ intermetallic layer, whose fracture elongation is even higher than that of the constitute layers. Guo et al. [74] reported that the intermetallic reacted layer can influent the fracture mechanism of adjacent layer. The plastic deformation behavior of adjacent Al layer can be constrained by the hard brittle intermetallic reacted Al_2Cu layer, leading to brittle fracture morphologies with crack propagating along the grain boundary of Al. Liu et al. [54] proposed that addition of thinner Al layer can enhance the tensile plasticity of cold-rolled Cu without losing strength by delaying onset of premature localized necking. Overall, the constrained interface can effectively improve the strain delocalization of soft layers and delay the formation of tunnel cracks [79].

4.5. Strain hardening exponent (n) and strain rate parameter (m)

For the strong interface without transition phase, the deformation behavior of laminated composites can depend on the strain hardening exponent (n) and strain rate parameter (m) of the constitute layers. Based on the power-law Hollomon equation, the flow stress (σ) of the individual layer can be expressed as follows [50]:

$$\sigma = K\left[\varepsilon^n + m\ln\left(\frac{\dot{\varepsilon}}{\dot{\varepsilon}_r}\right)\right] \tag{10}$$

where K is the strength constant and $\dot{\varepsilon}_r$ is a reference strain-rate. Herein, n and m exert the most dramatic influence, high value of n indicates the high uniform fracture elongation, and

Laminated systems	Brittle intermetallic layer	Maximum elongation at room temperature (%)
Ti-TiBw/Ti (present work)	–	27.5
Ti-Al [69]	Al3Ti, Al2Ti, TiAl, TiAl3	3.4
Ti-Cu [70]	Ti2Cu, TiCu, TiCu3, TiCu4	4.5
Ti-Ni [71]	TiNi3, Ti2Ni, TiNi	≤3
Ti-Nb [72]	NbTi	3.6
Ni-Al [73]	Ni3Al, Al2Al3, NiAl	12
Cu-Al [74]	Al2Cu, AlCu, Al3Cu4, Al2Cu3, Al4Cu9	11
Fe-Al [75, 76]	Fe2Al5, Fe4Al13	18
Fe-Cu [77]	—	60

Table 2. The elongation of laminated composites at room temperature in reported different system.

high value of m will even the material superplastic. The strain-rate hardening increases the elongation, which not only delays the onset of instability, but also by retarding the consequences of necking. The laminated composites with superior tensile ductility can be obtained by adjusting the values of n and m of constitute layer [80].

Figure 18 shows the fracture elongation map of many power-law hardening monolithic materials [80, 81]. It is interesting to note that a low ductility material laminated with another material that can coax the laminated composites toward higher elongation. For example, austenitic stainless steel and commercial purity iron make them attractive options as the ductility-enhancing component within a laminated metal composites, and this rule can also be indicated in the stainless steel/ultrahigh carbon steel multilayered materials during the high-temperature tensile testing process [32].

4.6. Tunnel crack

In 2008, Inoue et al. [37] reported that fracture elongation is gradually increased with the decrease of layer thickness, which is attributed to decreasing size of tunnel crack. For the

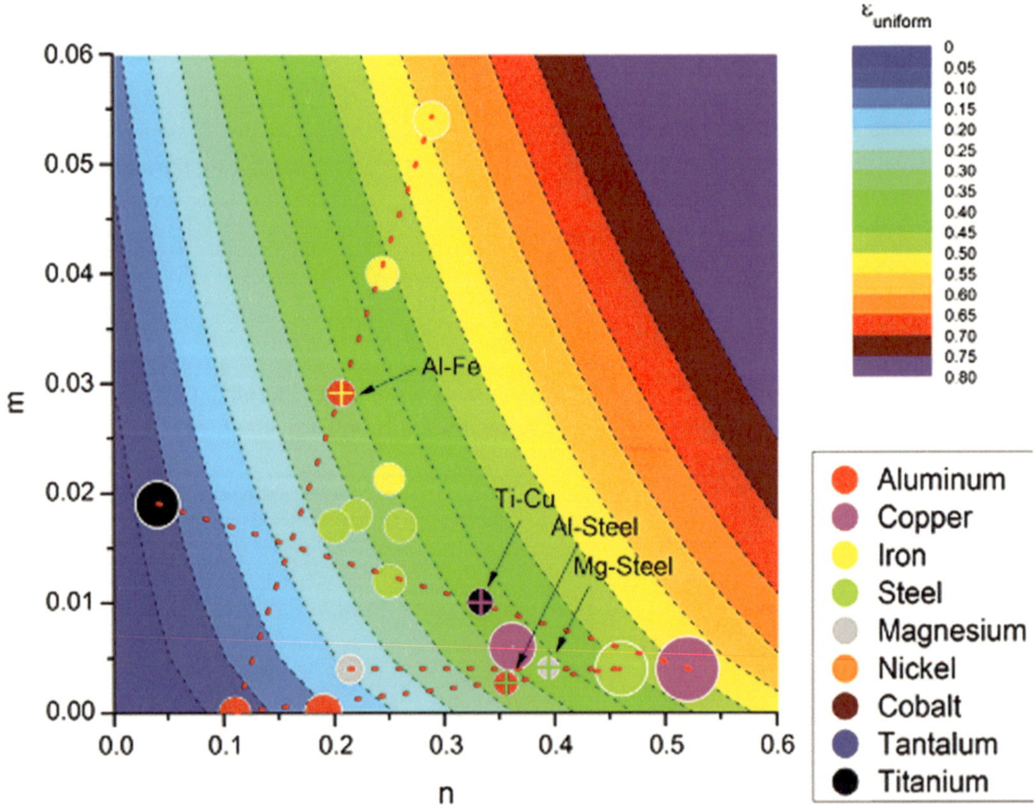

Figure 18. The fracture elongation map of many power-law hardening monolithic materials [80].

elastic-plastic laminated composites, the following criterion has already been proposed for suppressing crack propagation [40]:

$$t_B \leq t_{critical} = \frac{2\sqrt{3}K_{IC}^2\sigma_{Y,D}}{\sigma^3} \tag{11}$$

where t_B, K_{IC} and σ are the thickness, fracture toughness and the ultimate strength of the brittle phase layer, respectively. $\sigma_{Y,D}$ is the yielding strength of ductile phase layer, and $t_{critical}$ is the critical size. The above equation indicated that if the thickness brittle phase layer is lower than the critical thickness size, the tunnel crack in brittle phase layer can be effectively suppressed by the ductile phase layer. Therefore, a notable transition in the fracture morphologies of S420 brittle layer from brittle cleavage to ductile shear and dimples is identified as the layer thickness is reduced.

Meanwhile, Liu et al. [65, 82] reported that the laminated Ti-TiBw/Ti composites with thinner layer thickness possess higher fracture elongation than the laminated composites with thick layer thickness due to presence of more tunnel cracks. **Figure 19** shows the in situ tensile fracture process of laminated Ti-TiBw/Ti composites with layer thickness of 500 and 300 μm. It is revealed that the laminated composites with 500 μm layer thickness sharply fail at the tensile strain of 12.7%, and only a tunnel crack propagates into the overall sample as shown in **Figure 19(f)**. That is to say, when the brittle layer thickness extends beyond the critical thickness size, the ductile layer cannot prevent/blunt the propagation of tunnel crack, and the laminated composites will sharply fracture. However, the laminated composites with layer thickness of 300 μm revealed that one tunnel crack is presented at the strain of 13.4% as shown in **Figure 19(k)**. Finally, the laminated composites fail with two or more tunnel cracks at the strain of 15.4% as shown in **Figure 19(l)**, which is taken as a noncatastrophic fracture stage.

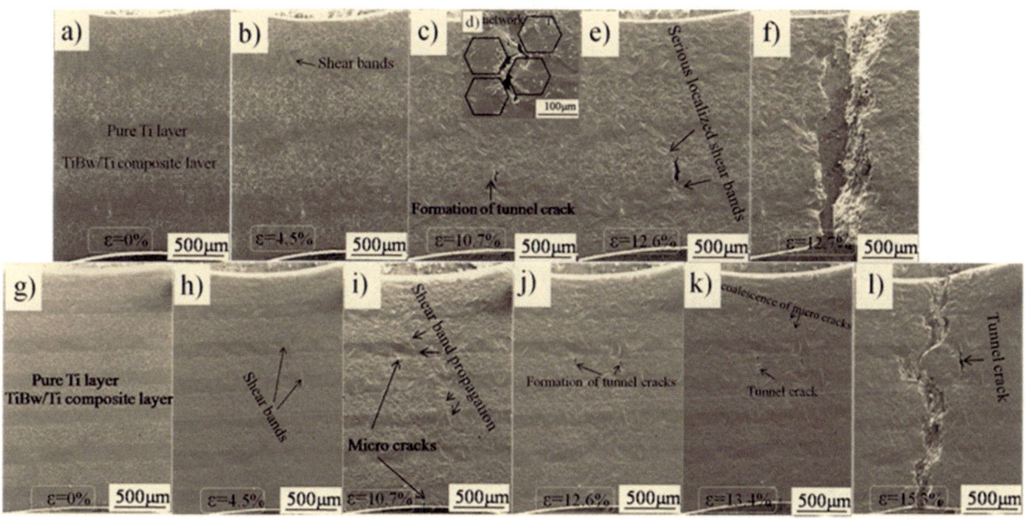

Figure 19. The fracture processes of the laminated Ti-TiBw/Ti composites with different layer thicknesses at the in situ tensile strains of (a, f) 0%; (b, h) 4.5%; (c, d, i) 10.7%; (e, j) 12.6%; (f) 12.7%; (k) 13.4%; (l) 15.3%. (a–f) For the laminated composites with 500 μm layer thickness; (g–l) for the laminated composites with 300 μm layer thickness [65].

Therefore, the laminated composites with 300 μm layer thickness can obtain higher fracture elongation than the composites with 500 μm layer thickness, which is attributed to more tunnel cracks formed in the thin TiBw/Ti layers.

In the laminated Ti-TiBw/Ti systems, Liu et al. reported [47, 83–87] that the formation and number of tunnel crack are also related to interfacial bonding strength and residual stress, and the ductile Ti layer exhibits ductile-brittle transition at the tunnel crack with the decreasing Ti layer thickness and the increasing volume fraction of TiB reinforcement, which is attributed to that the plastic deformation of thin Ti layer is constrained by TiBw/Ti layer, resulting in high triaxial tensile stress at the tunnel crack. Therefore, the laminated composites with thin Ti layer thickness reveal brittle fracture characteristics. With increasing volume fraction from 5 to 12%, the fracture manner in the TiBw/Ti layer changes from quasi-cleavage fracture to flat brittle fracture. Moreover, Ti layer reveals obviously ductile-brittle transition manner: the ductile dimples, crack bifurcation and lath-like intergranular fracture are presented in the laminated composites with volume fraction of 5%, 8% and 10%, respectively, and a typical brittle fracture manner is existed in the Ti layer. Huang et al. [85] reported that the plastic deformation zone can effectively delay the propagating behavior of tunnel crack. However, the size of plastic zone is decreased with the increasing volume fraction of TiB. Therefore, the tunnel crack is difficult to be blunted and easy to propagate in a brittle manner in the laminated composites with high volume fraction of reinforcements.

5. Conclusions

A background to the development of toughening mechanisms of metal matrix composites with multiscale hierarchical structure and laminated structure was overviewed in this article as follows:

1. The development of metal matrix composites approaches to the microstructure configuration design, and titanium matrix composites can achieve both high strength and toughness by adjusting the multiscale hierarchical structure.

2. The design ideal of metal matrix composites with laminated structure is aiming to achieve high strength, toughness and ductility. It is revealed that elastic property, and yield strength of laminated composites follow the "rule of average." However, the estimation of fracture elongation and fracture toughness is complex, which is inconsistent with the "rule of average."

3. The fracture elongation and toughness of laminated composites are related to interfacial bonding strength and tunnel crack number, and the fracture elongation of laminated composites with strong interface can be controlled by adjusting the strain hardening exponent n and strain rate parameter m.

4. The tensile properties and fracture characteristics of laminated composites reveal obvious size effect, with the decrease of layer thickness; the tensile strength is gradually improved, whereas the fracture elongation is first increased in microscale and then sharply decreased in the submicroscale and nanoscale.

Acknowledgements

This work is financially supported by the National Natural Science Foundation of China (NSFC) under Grant Nos. 51601055, 51471063, 51571071, 51671068, the Hebei Science and Technology program under Grant No. 130000048 and the National Natural Science Foundation of Hebei Province under Grant Nos. E201620218 and QN2016029.

Author details

Baoxi Liu[1], Lujun Huang[2]*, Lin Geng[2] and Fuxing Yin[1]

*Address all correspondence to: huanglujun@hit.edu.cn

1 Tianjin Key Laboratory of Materials Laminating Fabrication and Interfacial Controlling Technology, School of Materials Science and Engineering, Research Institute for Energy Equipment Materials, Hebei University of Technology, Tianjin, China

2 School of Materials Science and Engineering, Harbin Institute of Technology, Harbin, China

References

[1] Tjong C, Ma ZY. Microstructural and mechanical characteristics of in situ metal matrix composites. Materials Science and Engineering R. 2000;**29**:49-113

[2] Huang LJ, Geng L, Peng HX. Microstructurally inhomogeneous composites: Is a homogeneous reinforcement distribution optimal? Progress in Materials Science. 2005;**71**:93-168

[3] Tjong SC, Mai YW. Processing-structure-property aspects of particulate- and whisker-reinforced titanium matrix composites. Composites Science and Technology. 2008;**68**:583-601

[4] Lu K. Making strong nanomaterials ductile with gradients. Science. 2014;**345**:1455-1456

[5] Barthelat F. Biomimetics for next generation materials. Philosophical Transactions of the Royal Society A. 2007;**365**:2907-2919

[6] Lu K. The future of metals. Science. 2010;**328**:319-320

[7] Fang TH, Li WL, Tao NR, Lu K. Revealing extraordinary intrinsic tensile plasticity in gradient nano-grained copper. Science. 2011;**331**:1587-1590

[8] Liu XC, Zhang HW, Lu K. Strain-induced ultrahard and ultrastable nanolaminated structure in nickel. Science. 2013;**342**:337-340

[9] Palkovic SD, Brommer DB, Kupwade-Patil K, Masic A, Buehler MJ, Buyukozturk O. Roadmap across the mesoscale for durable and sustainable cement paste: A bioinspired approach. Construction and Building Materials. 2016;**115**:13-31

[10] Zheng XY, Smith W, Jackson J, Moran B, Cui H, Chen D, Ye JC, Fang N, Rodriguez N, Weisgraber T, Spadaccini CM. Multiscale metallic metamaterials. Nature Materials. 2016;**15**:1100-1106

[11] Yin L. Composites microstructures with tailored phase contiguity and spatial distribution. PHD thesis of University of Bristol; 2009. p. 1-25

[12] Radhakrishnabhat BV, Subramanyam J, Bhanuprasad VV. Preparation of Ti-TiB-TiC & Ti-TiB composites by in-situ reaction hot pressing. Materials Science and Engineering A. 2002;**325**:126-130

[13] Morsi K, Patel VV. Processing and properties of titanium-titanium boride dual matrix composites. Journal of Materials Science. 2007;**42**:2037-2047

[14] Morsi K, Patel VV, Nraghi S. Processing of titanium-titanium boride dual matrix composites. Journal of Materials Processing Technology. 2008;**196**:236-242

[15] Jiao Y, Huang LJ, Duan TB, Wei SL, Kaveendran B, Geng L. Controllable two-scale network architecture and enhanced mechanical properties of (Ti_5Si_3+TiBw)/Ti_6Al_4V composites. Scientific Reports. 2016;**6**:32991

[16] Jiao Y, Huang LJ, Wang S, Li XT, An A, Cui XP, Geng L. Effects of first-scale TiBw on secondary-scale Ti_5Si_3 characteristics and mechanical properties of in-situ (Ti_5Si_3+TiBw)/Ti_6Al_4V composites. Journal of Alloys and Compounds. 2017;**704**:269-281

[17] Duan HQ, Han YF, Lv WJ, Wang LQ, Mao JW, Zhang D. Configuration design and fabrication of laminated titanium matrix composites. Materials and Design. 2016;**99**: 219-224

[18] Panda KB, Ravichandran KS. Titanium-titanium boride (Ti-TiB) functionally graded materials through reaction sintering: Synthesis, microstructure, and properties. Metallurgical and Materials Transactions A. 2003;**34**:1994-2003

[19] Pettersson A, Magnusson P, Lundberg P, Nygren M. Titanium–titanium diboride composites as part of a gradient armour material. International Journal of Impact Engineering. 2005;**32**:387-399

[20] Zhang ZH, Shen XB, Zhang C, Wei S, Lee S, Wang FC. A new rapid route to in-situ synthesize TiB–Ti system functionally graded materials using spark plasma sintering method. Materials Science & Engineering A. 2013;**565**:326-332

[21] Ma J, Tan GEB. Processing and characterization of metal-ceramics functionally gradient materials. Journal of Materials Processing Technology. 2001;**113**:446-449

[22] Agrawal P, Sun CT. Fracture in metal-ceramic composite. Composites Science and Technology. 2004;**64**:1167-1178

[23] Huang LJ, Geng L, Wang B, Xu HY, Kaveendran B. Effects of extrusion and heat treatment on the microstructure and tensile properties of in situ TiBw/Ti_6Al_4V composite with a network architecture. Composites Part A. 2012;**43**:486-491

[24] Liu BX, Huang LJ, Geng L. Elastic and plastic behaviors of laminated Ti-TiBw/Ti composites. Journal of Wuhan University of Technology-Materials Science Edition. 2015;**30**: 596–600

[25] Peng HX. A review of "consolidation effects on tensile properties of an elemental Al matrix composite". Materials Science and Engineering A. 2005;**396**:1-6

[26] Hashin Z, Shtrikman S. A variational approach to the theory of the elastic behavior of polycrystals. Journal of the Mechanics and Physics of Solids. 1965;**38**:343-352

[27] Huang LJ, Geng L, Fu Y, Kaveendran B, Peng HX. Oxidation behavior of in situ TiCp/ Ti_6Al_4V composite with self-assembled network microstructure fabricated by reaction hot pressing. Corrosion Science. 2013;**69**:175-180

[28] Lin T, Evans AG, Ritchie RO. Stochastic modeling of the independent roles of particle size and grain size in transgranular cleavage fracture. Metallurgical Transactions A. 1987;**18**:641-651

[29] Ma J, Tan GEB, He Z. Fabrication and characterization of Ti-TiB_2 functionally graded material system. Metallurgical and Materials Transactions A. 2002;**33**:681-685

[30] Min XH, Emura S, Meng FQ, Mi GB, Tsuchiya K. Mechanical twinning and dislocation slip multilayered deformation microstructures in b-type Ti–Mo base alloy. Scripta Materialia. 2015;**102**:79-82

[31] Gutierrez-Urrutia I, Li CL, Emura S, Min XH, Tsuchiya K. Study of {332}<113> twinning in a multilayered Ti-10Mo-xFe (x = 1–3) alloy by ECCI and EBSD. Science and Technology of Advanced Materials. 2016;**17**:221-228

[32] Lesuer DR, Syn CK, Sherby OD, Wadsworth J, Lewandowski JJ, Hunt WH. Mechanical behavior of laminated metal composites. International Materials Reviews. 1996;**41**:169-197

[33] Wadsworth J, Lesuer DR. Ancient and modern laminated composites: From the great Pyramid of Gizeh to Y2K. Materials Characterization. 2000;**45**:289-313

[34] Immarigeon JP, Holt RT, Koul AK, Zhao L, Wallace W, Beddoes JC. Lightweight materials for aircraft applications. Materials Characterization. 1995;**35**:393-399

[35] Price RD, Jiang FC, Kulin RM, Vecchio KS. Effect of ductile phase volume fraction on the mechanical properties of Ti-Al_3Ti metal-intermetallic laminate (MIL) composites. Materials Science and Engineering A. 2011;**528**:3134-3146

[36] Yin FX, Li L, Tanaka Y, Kishimoto S, Nagai K. Hot rolling bonded multilayered composite steels and varied tensile deformation behavior. Materials Science and Technology. 2012;**28**:783-787

[37] Inoue J, Nambu S, Ishimoto Y, Koseki T. Fracture elongation of brittle/ductile multilayered steel composites with a strong interface. Scripta Materialia. 2008;**59**:1055-1058

[38] Nambu S, Michiuchi M, Inoue J, Koseki T. Effect of interfacial bonding strength on the tensile ductility of multilayered steel composites. Composites Science and Technology. 2009;**69**:1936-1941

[39] Nambu S, Michiuchi M, Ishimoto Y, Asakura K, Inoue J, Koseki T. Transition in deformation behavior of martensitic steel during large deformation under uniaxial tensile loading. Scripta Materialia. 2009;**60**:221-224

[40] Koseki T, Inoue J, Nambu S. Development of multilayer steels for improved combinations of high strength and high ductility. Materials Transactions. 2014;**55**:227-237

[41] Kimura Y, Inoue T, Yin FX, Tsuzaki K. Inverse temperature dependence of toughness in an ultrafine grain-structure steel. Science. 2008;**320**:1057-1059

[42] Wu H, Fan GH, Huang M, Geng L, Cui XP, Xie HL. Deformation behavior of brittle/ductile multilayered composites under interface constraint effect. International Journal of Plasticity. 2016;**89**:96-109

[43] Zhou P, Hu P, Zhang XH, Han WB. Laminated ZrB_2-SiC ceramic with improved strength and toughness. Scripta Materialia. 2011;**64**:276-279

[44] Go YH, Cho J, Jeong CY, Kang CS, Park JS. Stress distribution of bulk metallic glass/metal laminate composites during uniaxial fracture. Materials Science and Engineering A. 2007;**460-461**:377-382

[45] Hwu KL, Derby B. Fracture of metal/ceramic laminates-I. Transition from single to multiple cracking. Acta Materialia. 1999;**47**:529-543

[46] Bouaziz O, Masse JP, Petitgand G, Huang MX. A novel strong and ductile TWIP/martensite steel composite. Advanced Engineering Materials. 2016;**18**:56-59

[47] Liu BX, Huang LJ, Rong XD, Geng L, Yin FX. Bending behaviors and fracture characteristics of laminated ductile-tough composites under different modes. Composites Science and Technology. 2016;**126**:94-105

[48] Rohatgi A, Harach DJ, Vecchio KS, Harvey KP. Resistance-curve and fracture behavior of Ti-Al_3Ti metal-intermetallic laminate (MIL) composites. Acta Materialia. 2003;**51**: 1933-2957

[49] Lugovy M, Slyunyayev V, Orlovskaya N, Blugan G, Kuebler J, Lewis M. Apparent fracture toughness of Si_3N_4-based laminates with residual compressive or tensile stresses in surface layers. Acta Materialia. 2005;**53**:289-296

[50] Cetin A, Bernardi C, Mortensen A. An analysis of the tensile elongation to failure of laminated metal composites in the presence of strain-rate hardening. Acta Materialia. 2012;**60**:2265-2276

[51] Hutchinson JW, Neale KW. Influence of strain-rate sensitivity on necking under uniaxial tension. Acta Materialia. 1977;**25**:839-846

[52] Hutchinson JW. Necking modes in multilayers and their influence on tearing toughness. Mathematics and Mechanics of Solids. 2014;**19**:39-55

[53] Serror MH. Analytical study for deformability of laminated sheet metal. Journal of Advanced Research. 2013;**4**:83-92

[54] Liu HS, Zhang B, Zhang GP. Delaying premature local necking of high strength Cu: A potential way to enhance plasticity. Scripta Materialia. 2011;**64**:13-16

[55] Li YP, Zhu XF, Zhang GP, Tan J, Wang W, Wu B. Investigation of deformation instability of Au/Cu multilayers by indentation. Philosophical Magazine. 2010;**90**:3049-3067

[56] Chen Z, Mecholsky JJ. Control of strength and toughness of ceramic/metal laminates using interface design. Journal of Materials Research. 1993;**8**:2362-2369

[57] Zhang ZL. The research on the processing and mechanical properties of metal-ceramics laminated ceramic [thesis]. Harbin Institute of Technology; 2010 [in Chinese]. pp. 1-55. http://epub.cnki.net/kns/brief/result.aspx?dbPrefix=CDMD

[58] Cook J, Gordon JE, Evans CC, Marsh DM. A mechanism for the control of crack propagation in all-brittle systems. Proceedings of the Royal Society of London. Series A. 1964;**282**:508-520

[59] Jin ZH, Paulino GH, Dodds RH. Cohesive fracture modeling of elastic-plastic crack growth in functionally graded materials. Engineering Fracture Mechanics. 2003;**70**:885-1912

[60] Launey ME, Ritchie RO. On the fracture toughness of advanced materials. Advanced Materials. 2009;**21**:2103-2110

[61] Li TZ, Grignon F, Benson DJ, Vecchio KS, Olevsky EA, Jiang FC, Meyers MA. Modeling the elastic properties and damage evolution in Ti-Al_3Ti metal-intermetallic laminate (MIL) composites. Materials Science and Engineering A. 2004;**374**:10-26

[62] Liu BX, Geng L, Dai XL, Yin FX, Huang LJ. Multiple toughening mechanisms of laminated Ti-TiBw/Ti composites fabricated by diffusion welding. Materials Science Forum. 2016;**848**:196-201

[63] Xu Q, Hayes RW, Huntjr WH, Lavernia EJ. Mechanical properties and fracture behavior of layered 6061/SiCp composites produced by spray atomization and co-deposition. Acta Materialia. 1999;**47**:43-53

[64] Hidane A, Shukla A. Quasi-static and dynamic fracture initiation toughness of Ti/TiB layered functionally graded material under thermo-mechanical loading. Engineering Fracture Mechanics. 2010;**77**:479-491

[65] Liu BX, Huang LJ, Geng L, Wang B, Cui XP. Fracture behaviors and microstructural failure mechanisms of laminated Ti-TiBw/Ti composites. Materials Science and Engineering A. 2014;**611**:290-297

[66] Gao HJ, Ji BH, Jager IL, Arzt E, Fratzi P. Materials become insensitive to flaws at nanoscale: Lessons from nature. Proceedings of the National Academy of Sciences of the United States of America. 2003;**10**:5597-5600

[67] Hsia KJ, Suo Z, Yang W. Cleavage due to dislocation confinement in layered materials. Journal of the Mechanics and Physics of Solids. 1994;**42**:877-896

[68] Li T, Suo Z. Deformability of thin metal films on elastomer substrates. International Journal of Solids and Structures. 2006;**43**:2351-2363

[69] Mizuuchi K, Inoue K, Sugioka M, Itami M, Kawahara M. Microstructure and mechanical properties of Ti-aluminides reinforced Ti matrix composites synthesized by pulsed current hot pressing. Materials Science and Engineering A. 2004;**368**:260-268

[70] Konieczny M, Dziadon A. Strain behavior of copper-intermetallic layered composite. Materials Science and Engineering A. 2007;**460-461**:238-242

[71] Bunshah RF, Nimmagadda R, Doerr HJ. Structure-property relationships in Cr/Cu and Ti/Ni microlaminate composites. Thin Solid Films. 1984;**112**:227-236

[72] Rudnitskii NP. Study on the strength and plasticity of Ti-Nb-based laminated composites in a temperature range of 290-1700K. Strength of Materials. 2002;**34**:612-616

[73] Konieczny M. Microstructural characterization and mechanical response of laminated Ni-intermetallic composites synthesized using Ni sheets and Al foils. Materials Characterization. 2012;**70**:117-124

[74] Guo YJ, Qian GJ, Jian WZ, Zhi XH. Microstructure and tensile behavior of Cu-Al multilayered composites prepared by plasma activated sintering. Materials Science and Engineering A. 2010;**50**:5234-5240

[75] Talebian M, Alizadeh M. Manufacturing Al/steel multilayered composite by accumulative roll bonding and the effects of subsequent annealing on the microstructural and mechanical characteristics. Materials Science and Engineering A. 2014;**590**:186-193

[76] Wang Y, Vecchio KS. Microstructure evolution in Fe-based-aluminide metallic-intermetallic laminate (MIL) composites. Materials Science and Engineering A. 2016;**649**:325-337

[77] Tsai HC, Wolfenstine J, Sherby OD. Elevated temperature mechanical behavior of an ultrahigh carbon steel/brass laminated composites. Composites. 1991;**22**:373-379

[78] Pang JC, Fan GH, Cui XP, Li AB, Geng L, Zheng ZZ, Wang QW. Mechanical properties of Ti-(SiCp/Al) laminated composite with nano-sized $TiAl_3$ interfacial layer synthesized by roll bonding. Materials Science and Engineering A. 2013;**582**:294-298

[79] Huang M, Fan GH, Geng L, Cao GJ, Du Y, Wu H, Zhang TT, Kang HJ, Wang TM, Du GH, Xie HL. Revealing extraordinary tensile plasticity in layered Ti-Al metal composite. Scientific Reports. 2016;**6**:1-10

[80] Cohades A, Cetin A, Mortensen A. Designing laminated metal composites for tensile ductility. Materials and Design. 2015;**66**:412-420

[81] Cohades A, Mortensen A. Tensile elongation of unidirectional or laminated composites combining a brittle reinforcement with a ductile strain and strain-rate hardening matrix. Acta Materialia. 2014;**71**:31-43

[82] Liu BX. Research on fabrication and mechanical behavior of laminated Ti-TiBw/Ti composites [thesis]. Harbin Institute of Technology; 2014. pp. 1-140 http://epub.cnki.net/kns/brief/result.aspx?dbPrefix=CDMD

[83] Liu BX, Huang LJ, Wang B, Geng L. Effect of pure Ti thickness on the tensile behavior of laminated Ti-TiBw/Ti composites. Materials Science and Engineering A. 2014;**617**:115-120

[84] Liu BX, Huang LJ, Geng L, Wang B, Cui XP. Effects of reinforcement volume fraction on tensile behaviors of laminated Ti-TiBw/Ti composites. Materials Science and Engineering A. 2014;**610**:344-349

[85] Huang LJ, Geng L, Wang B, Wu LZ. Effects of volume fraction on the microstructure and tensile properties of in situ TiBw/Ti_6Al_4V composites with novel network microstructure. Materials and Design. 2013;**45**:532-538

[86] Liu BX, Huang LJ, Geng L, Kaveendran B, Wang B, Song XQ, Cui XP. Gradient grain distribution and enhanced properties of novel laminated Ti-TiBw/Ti composites by reaction hot-pressing. Materials Science and Engineering A. 2014;**595**:257-265

[87] Liu BX, Huang LJ, Kaveendran B, Geng L, Cui XP, Wei SL, Yin FX. Tensile and bending behaviors and characteristics of laminated Ti-(TiBw/Ti) composites with different interface status. Composites Part B. 2017;**108**:377-385

Chapter 3

Bending of Laminated Composite Plates in Layerwise Theory

Marina Rakočević

Additional information is available at the end of the chapter

http://dx.doi.org/10.5772/intechopen.69975

Abstract

Determination of stress-strain state in contemporary laminated composite plates containing layers with continuous unidirectional fibers requires the application of refined plate theories, which include layerwise theory. In contrast to homogeneous isotropic plates, heterogeneity of the anisotropic structure of laminated composite plates often leads to the appearance of imperfections in the connection between the layers. Mathematical models, which are formed on the assumption that the plate is homogeneous and isotropic, cannot properly include irregularities that can occur at the level of the layer in the process of manufacture, transportation, installation, or exploitation. Mathematical models of layerwise theory allow defining a more realistic stress-strain state through the thickness of the plate, where consideration is carried out at the level of the layer. Additionally, this model makes possible to include delaminations that might occur on the connection between the individual layers. In this chapter, Reddy's layerwise theory is applied in order to determine equations for the problem of bending of laminated composite plates. The bending equations are solved by applying analytical method by means of double trigonometric series, as well as by using numerical methods based on the finite elements. This chapter presents examples for both applied approaches.

Keywords: laminated plate, bending, layerwise theory, analytical solution, finite element method

1. Introduction

Composite materials are obtained by a combination of two or more materials which together have characteristics that separately usually cannot be achieved with reasonable costs. During the last decades, a sudden increase of the use of modern composite materials is based on the achieved advanced properties which include their lower weight and higher strength and resistance compared to the classical and conventional composite materials. It is clear that the modern composite materials are materials of the future, because they allow the designers to choose the appropriate characteristics of the elements structure depending on the applicable problem. Due to its complexity, the researches of modern materials involve several scientific fields, and each of them, in their own way, contributes to their development.

Composite plates, made of layers of the various material and geometric characteristics, are called laminated plates. Ply or layer is the basic element of the laminated plate and it is made of fibers installed into the matrix. The fibers can be discontinuous, continuous unidirectional, bidirectional, woven unidirectional, or woven bidirectional. In the unidirectional laminated composites, the fibers are load-bearing elements of the layer, while the matrix has a role to protect the fibers from external influences, to hold the fibers together and to perform uniform distribution of the influences to each of the fibers. The materials used for fibers have better material property and greater capacity compared to the matrix, and the geometrical characteristics of the fiber cross section are significantly smaller than its length. Materials used for fibers can be aluminum, copper, iron, nickel, steel, titanium, or organic materials such as glass, carbon, and graphite. A layer with unidirectional fibers has significantly better characteristics in fiber direction than in a direction perpendicular to the fiber.

Heterogeneity of anisotropic laminated composite plates often causes the appearance of a large number of imperfections that can occur at the level of the laminated plate or at the level of the layer, as well as at the local level of the fiber/matrix. General deformation of laminated plates is often defined by complex coupling between the axial deformation, bending, and shear deformation. In laminated composite plates for smaller aspect ratio, the importance of shear deformation is higher than in the corresponding homogeneous isotropic plates.

At the level of the layer, composites often contain concentrations of transverse stresses near the geometrical and material imperfections which lead to damage. Discontinuity between adjacent layers can occur in the stages of production, transportation, installation, or exploitation. This phenomenon in the literature is known as delamination. Delamination and its increase during exploitation is one of the most important parameters that affect the bearing capacity of the plate. Plates with delamination have reduced bearing capacity to bending. In the analysis of these plates, it is necessary to choose the mathematical models which can successfully incorporate delamination in the calculations.

Mathematical models for these particular problems need to determine the real stress-strain state in the laminated plate, which requires the application of more accurate theories. In addition, it is important to find a balance between the desired accuracy and calculation costs.

The literature that studies the problems of laminated composite plates is significant and extensive. Available mathematical models for the laminated plates are based on assumptions of Equivalent Single Layer (ESL) theories and assumptions of LayerWise (LW) theories, see Refs. [1–26]. ESL theories consider plate as one layer with equivalent stiffnesses. The most commonly applied ESL theories of laminated composite plates are Classical Laminated Plate Theory (CLPT), First-Order Shear Deformation Theory (FSDT), Third-Order Shear Deformation Theory (TSDT), and other high-order shear theories. ESL theories are recommended to apply for thin and very thin laminated plates. More accurate ESL plate theories introduce shear deformation in the calculation which increases accuracy compared to the classical laminated theory. Layerwise theories consider laminated composite plate at the level of the layer. LW theories are recommended for moderately thick and thick laminated plates and for plates with a pronounced anisotropic behavior for which it is necessary to define a more realistic internal stresses distribution, as well as where it is necessary to include delamination and the other geometrical and material imperfections in calculation. On the basis of LW theories, many calculation procedures, analytical, and numerical methods were created in order to accurately analyze mentioned stresses of the laminated composite plates.

This chapter describes the application of layerwise theories in solving problems of bending of laminated plates, whereby the three-dimensional (3D) problem of elasticity is reduced by solving differential equations with two-dimensional (2D) variables in the plane plate. Additionally, the approximation of the displacement through the thickness of the plate is conducted with one-dimensional interpolation functions of coordinate perpendicular to the plane of the layer. Mathematical model is based on the Generalized Laminated Plate Theory (GLPT) developed by Reddy and Reddy et al. [10–14].

Based on the adopted assumptions of the layerwise theory, bending equations are obtained by applying the principle of virtual displacements. In this chapter, the governing differential equations of LW theory are solved by applying analytical and numerical methods. Closed analytical solution of bending equations in LW theory is done by using the double trigonometric series of the simple supported rectangular plate with the angle of the layers orientation 0 and 90°, see Refs. [10, 18, 22–26].

Analytical solution for this mechanical problem is difficult to determine in case of complex plate geometry, arbitrary angles of layers orientation, arbitrary boundary conditions, as well as for various forms of nonlinearity. These are the reasons why the applied Finite Element Method (FEM) as one of the most used numerical methods was used, see Refs. [1, 10–14, 21]. The idea of the FEM is a representation of the domain as the sets of appropriate simple geometrical shapes called the finite elements. Discretization of the domain with finite elements makes this method a valuable tool in solving a large number of complex engineering problems.

Since the FEM is a method of discretization, errors can be classified into two categories: geometric discretization errors and interpolation errors. Accuracy of the results obtained by applying FEM depends on the choice of interpolation function in the plane (x, y) and interpolation function through the thickness of the plate. It also depends on the number of elements and the number of nodes of the adopted network in the plane plate and perpendicular to the plane plate. The increase of the number of finite element nodes gives rise to the increase of the number of degrees of freedom, making this calculation uneconomic. Regardless of the pointed disadvantages, FEM is much more general and more acceptable for the analysis of laminated composite plates than analytical methods. In this chapter, the variable kinematic finite element (with C^0 continuity through the plate thickness) based on the assumptions of the partial LW theory is presented. The element is hierarchical and can include delamination. This finite element can be expanded in order to obtain a more complex finite element, by simply adding neglected elongation perpendicular to the plane of the plate.

2. Theoretical backgrounds

Laminated composite plate, loaded by transversal load, is considered in Cartesian coordinate system (x, y, z). The plate includes N orthotropic layers wherein each layer carries a predetermined direction in the plane of the plate (x, y). Layer orientation angle is defined by angle θ, angle between continuous fibers, and axis x of the global coordinate system of the plate.

It is assumed that the dilation ε_z perpendicular to the middle plane of the plate is neglected which results in a constant deflection through the plate thickness $w(x, y, z) = w(x, y)$. A displacement field is determined by components:

$$\begin{aligned} u(x,y,z) &= u(x,y) + U(x,y,z) \\ v(x,y,z) &= v(x,y) + V(x,y,z) \\ w(x,y,z) &= w(x,y) \end{aligned} \tag{1}$$

Displacement components $u(x, y, z)$ and $v(x, y, z)$ are obtained as a sum of displacements of the middle plane of the plate $u(x, y)$, $v(x, y)$ and the additional displacements $U(x, y, z)$ and $V(x, y, z)$ through the plate thickness. Reduction of a 3D problem to a 2D problem is carried out by introducing the following relations:

$$\begin{aligned} U(x,y,z) &= \sum_{J=1}^{n} u^J(x,y)\psi^J(z) \\ V(x,y,z) &= \sum_{J=1}^{n} v^J(x,y)\psi^J(z) \end{aligned} \tag{2}$$

where u^J and v^J are the displacement components of Jth node through the plate thickness in the directions of x- and y-axes, respectively; $\psi^J(z)$ is the one-dimensional (1D) interpolation function.

The governing equations of the laminated composite plates in case of bending are solved by displacement method of analysis, so that the primary variables are displacement components defined by Eqs. (1) and (2). The primary variables can be conveniently displayed in a vector form:

$$\{\Delta\} = \begin{Bmatrix} u \\ v \\ w \end{Bmatrix} \quad \{\Delta^J\} = \begin{Bmatrix} u^J \\ v^J \end{Bmatrix} \tag{3}$$

where Δ are the components of the displacement vector of the middle plane plate and Δ^J are the components of the displacement vector of the Jth node through the plate thickness.

By relations (Eq. (2)), the discretization through the plate thickness is performed, wherein the u^J and v^J are unknown nodal displacements and $\psi^J(z)$ are adopted general 1D interpolation functions that can be polynomials. The degree of a polynomial is related to the number of nodes in which determined unknown nodal displacements occur, where

$$n = pN + 1 \tag{4}$$

where p is the polynomial degree; n is the number of nodes through the plate thickness; N is the number of layers.

For linear interpolation $p = 1$, the number of nodes through the plate thickness is one larger than the number of layers $n = N+1$, and the linear interpolation functions are determined by

$$\phi^J(z) = \left\{ \begin{array}{l} \psi_1^{J-1}(z) = \dfrac{z - z_{J-1}}{z_J - z_{J-1}}, z_{J-1} < z < z_J \\ \psi_2^J(z) = \dfrac{z_{J+1} - z}{z_J - z_{J-1}}, z_J < z < z_{J+1} \end{array} \right\} \tag{5}$$

where z_J is the coordinate of the Jth node which is located between the layers of J and J-1.

Graphical display of the laminated composite plates with linear interpolation through the plate thickness is shown in **Figure 1**. By analyzing separate layers, this LW theory can be understood as a first-order shear deformation theory for each layer of the observed laminated plate. A layer of the layered plates has an effect on the previous and the next layer of the laminate.

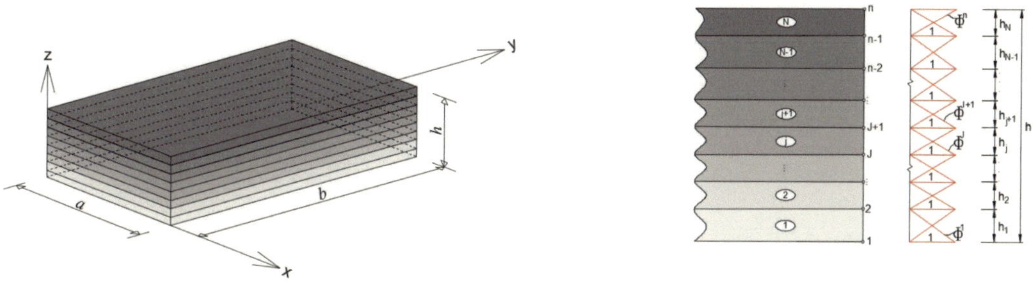

Figure 1. 3D geometry of the laminated composite plates and discretization through the plate thickness with linear interpolation.

Based on Eqs. (1) and (2), strain components can be expressed as follows:

$$\{\varepsilon\} = \left\{ \begin{array}{c} u_{,x} \\ v_{,y} \\ u_{,y} + v_{,x} \\ w_{,x} \\ w_{,y} \end{array} \right\} \quad \{\varepsilon^J\} = \left\{ \begin{array}{c} U^J_{,x} \\ V^J_{,y} \\ U^J_{,y} + V^J_{,x} \\ U^J \\ V^J \end{array} \right\} \tag{6}$$

where the members of the vector ε determine the deformation of the middle plane of the plate, and the members of the vector ε^J determine the deformation through the thickness of the plate. The short form of Eq. (3) is defined by

$$\{\varepsilon\} = [H]\{\Delta\}, \quad \{\varepsilon^J\} = [\overline{H}]\{\Delta^J\} \tag{7}$$

where members of matrix H and $\overline{H}$ are the first derivative of the interpolation functions with respect to the global coordinates x and y.

For an orthotropic material, the constitutive equations of the kth layer are determined by

$$\{\sigma\}^{(k)} = \begin{bmatrix} \overline{Q}_{11} & \overline{Q}_{12} & \overline{Q}_{16} & 0 & 0 \\ \overline{Q}_{12} & \overline{Q}_{22} & \overline{Q}_{26} & 0 & 0 \\ \overline{Q}_{16} & \overline{Q}_{26} & \overline{Q}_{66} & 0 & 0 \\ 0 & 0 & 0 & \overline{Q}_{44} & \overline{Q}_{45} \\ 0 & 0 & 0 & \overline{Q}_{45} & \overline{Q}_{55} \end{bmatrix}^{(k)} \{\varepsilon\} \tag{8}$$

In the previous equation, $\overline{Q}_{ij}$ are members of the transformed plane stress-reduced stiffnesses of kth layer which are obtained as follows; see Ref. [1]:

$$\begin{aligned}
\overline{Q}_{11} &= Q_{11}\cos^4\theta + 2(Q_{12} + 2Q_{66})\sin^2\theta\cos^2\theta + Q_{22}\sin^4\theta \\
\overline{Q}_{12} &= (Q_{11} + Q_{22} - 4Q_{66})\sin^2\theta\cos^2\theta + Q_{12}\left(\sin^4\theta + \cos^4\theta\right) \\
\overline{Q}_{22} &= Q_{11}\sin^4\theta + 2(Q_{12} + 2Q_{66})\sin^2\theta\cos^2\theta + Q_{22}\cos^4\theta \\
\overline{Q}_{16} &= (Q_{11} - Q_{12} - 2Q_{66})\cos^3\theta\sin\theta + (Q_{12} - Q_{22} + 2Q_{66})\sin^3\theta\cos\theta \\
\overline{Q}_{26} &= (Q_{11} - Q_{12} - 2Q_{66})\sin^3\theta\cos\theta + (Q_{12} - Q_{22} + 2Q_{66})\cos^3\theta\sin\theta \\
\overline{Q}_{66} &= (Q_{11} + Q_{22} - 2Q_{12} - 2Q_{66})\sin^2\theta\cos^2\theta + Q_{66}\left(\sin^4\theta + \cos^4\theta\right) \\
\overline{Q}_{44} &= Q_{44}\cos^2\theta + Q_{55}\sin^2\theta \\
\overline{Q}_{45} &= (Q_{55} - Q_{44})\cos\theta\sin\theta \\
\overline{Q}_{55} &= Q_{55}\cos^2\theta + Q_{44}\sin^2\theta
\end{aligned} \tag{9}$$

where θ is the angle between the fibers and the x-axis of the global coordinate system of the plate and Q_{ij} is the plane stress-reduced stiffnesses of kth layer (including materials constants).

Transverse stresses σ_{xz}^k and σ_{yz}^k obtained using Eq. (8) have a constant value along the kth layer. In order to determine a more realistic distribution of transverse shear stresses through the thickness approximate method that is given in Refs. [23, 26] can be used.

The internal forces in a cross section of the laminated plate are defined as integrals of stresses

$$\begin{aligned}
\left(N_x, N_y, N_{xy}, Q_x, Q_y\right) &= \int_0^h \left(\sigma_{xx}, \sigma_{yy}, \sigma_{xy}, \sigma_{xz}, \sigma_{yz}\right) dz \\
\left(N_x^J, N_y^J, N_{xy}^J\right) &= \int_0^h \left(\sigma_{xx}, \sigma_{yy}, \sigma_{xy}\right)\phi^J(z) dz \\
\left(Q_x^J, Q_y^J\right) &= \int_0^h \left(\sigma_{xz}, \sigma_{yz}\right)\phi^J_{,z} dz
\end{aligned} \tag{10}$$

where $\sigma_x, \sigma_y, \sigma_{xy}, \sigma_{xz}, \sigma_{yz}$ are stress components.

The internal forces can be expressed in terms of the displacements by inserting Eq. (6) into Eq. (7) and then in the integrals in Eq. (10):

$$\begin{aligned} \{N\} &= [A]\{\varepsilon\} + \sum_{J=1}^{n}\left[B^J\right]\left\{\varepsilon^J\right\} \\ \left\{N^J\right\} &= \left[B^J\right]\{\varepsilon\} + \sum_{J=1}^{n}\left[D^{IJ}\right]\left\{\varepsilon^J\right\} \end{aligned} \tag{11}$$

where $[A]$, $\left[B^J\right]$, and $\left[D^{IJ}\right]$ are stiffnesses of a laminated plate

$$A_{ij} = \sum_{J=1}^{N}\int_{z_k}^{z_{k+1}} \overline{Q}_{ij}^{(k)} dz, \qquad (i,j = 1,2,6,4,5) \tag{12}$$

$$\begin{aligned} B_{ij}^J &= \sum_{k=1}^{N}\int_{z_k}^{z_{k+1}} \overline{Q}_{ij}^{(k)} \psi^J dz \qquad (i,j = 1,2,6) \\ B_{ij}^J &= \sum_{k=1}^{N}\int_{z_k}^{z_{k+1}} \overline{Q}_{ij}^{(k)} \psi^J_{,z} dz \qquad (i,j = 4,5) \end{aligned} \tag{13}$$

$$\begin{aligned} D_{ij}^{JI} &= \sum_{k=}^{N}\int_{z_k}^{z_{k+1}} \overline{Q}_{ij}^{(k)} \psi^J \psi^I dz \qquad (i,j = 1,2,6) \\ D_{ij}^{JI} &= \sum_{k=1}^{N}\int_{z_k}^{z_{k+1}} \overline{Q}_{ij}^{(k)} \psi^J_{,z} \psi^I_{,z} dz \qquad (i,j = 4,5) \end{aligned} \tag{14}$$

For the adopted linear interpolation functions (Eq. (5)) from Eqs. (12)–(14) can be obtained

$$\begin{aligned}
& A_{ij} = \sum_{k=1}^{N}\overline{Q}_{ij}^{K}(z_{k-1} - z_k), && (i,j = 1,2,6,4,5) \\
& B_{ij}^J = \overline{Q}_{ij}^{J-1}\frac{(z_J - z_{J-1})}{2} + \overline{Q}_{ij}^{J}\frac{(z_{J+1} - z_J)}{2} && (i,j = 1,2,6) \\
& B_{ij}^J = \overline{Q}_{ij}^{J-1} - \overline{Q}_{ij}^{J} && (i,j = 4,5) \\
& for \quad J = I && \\
& D_{ij}^{IJ} = \overline{Q}_{ij}^{J-1}\frac{(z_J - z_{J-1})}{3} + \overline{Q}_{ij}^{J}\frac{(z_{J+1} - z_J)}{3} && (i,j = 1,2,6) \\
& D_{ij}^{IJ} = \frac{\overline{Q}_{ij}^{J-1}}{(z_J - z_{J-1})} + \frac{\overline{Q}_{ij}^{J}}{(z_{J+1} - z_J)} && (i,j = 4,5) \\
& for \quad J = I \pm 1 && \\
& D_{ij}^{IJ} = D_{ij}^{JI} = \overline{Q}_{ij}^{J}\frac{(z_{J+1} - z_J)}{6} && (i,j = 1,2,6) \\
& D_{ij}^{IJ} = D_{ij}^{JI} = -\frac{\overline{Q}_{ij}^{J}}{(z_{J+1} - z_J)} && (i,j = 4,5)
\end{aligned} \tag{15}$$

It is noted that the elements of matrix A are determined for the laminated plate as a whole and do not depend on the adopted discretization through the thickness of the plate.

The governing equations of motion can be derived by using the principle of displacements. For zero values of stiffnesses $A_{16} = A_{26} = A_{45} = B_{16}^{J} = B_{26}^{J} = B_{45}^{J} = D_{16}^{JI} = D_{26}^{JI} = D_{45}^{JI}$, the equations take the form:

$$
\begin{aligned}
&A_{11}u_{,xx} + A_{12}v_{,yx} + A_{66}\left(u_{,yy} + v_{,xy}\right) + \sum_{J=1}^{n}\left[B_{11}^{J}U^{J}_{,xx} + B_{12}^{J}V^{J}_{,yx} + B_{66}^{J}\left(U^{J}_{,yy} + V^{J}_{,xy}\right)\right] = 0 \\
&A_{12}u_{,xy} + A_{22}v_{,yy} + A_{66}\left(u_{,yx} + v_{,xx}\right) + \sum_{J=1}^{n}\left[B_{12}^{J}U^{J}_{,xy} + B_{22}^{J}V^{J}_{,yy} + B_{66}^{J}\left(U^{J}_{,yx} + V^{J}_{,xx}\right)\right] = 0 \\
&A_{55}w_{,xx} + A_{44}w_{,yy} + \sum_{J=1}^{n}\left[B_{55}^{J}U^{J}_{,x} + B_{44}^{J}V^{J}_{,y}\right] + q = 0 \\
&B_{11}^{J}u_{,xx} + B_{12}^{J}v_{,xy} + B_{66}^{J}\left(u_{,yy} + v_{,xy}\right) - B_{55}^{J}w_{,x} + \sum_{J=1}^{n}\left[D_{11}^{JI}U^{J}_{,xx} + D_{12}^{JI}V^{J}_{,yx} + D_{66}^{JI}\left(U^{J}_{,yy} + V^{J}_{,xy}\right) - D_{55}^{JI}U^{I}\right] = 0 \\
&B_{11}^{J}u_{,xy} + B_{21}^{J}v_{,yy} + B_{66}^{J}\left(u_{,yx} + v_{,xx}\right) - B_{44}^{J}w_{,y} + \sum_{J=1}^{n}\left[D_{12}^{JI}U^{J}_{,xy} + D_{22}^{JI}V^{J}_{,yy} + D_{66}^{JI}\left(U^{J}_{,yx} + V^{J}_{,xx}\right) - D_{44}^{JI}V^{I}\right] = 0
\end{aligned}
\tag{16}
$$

where q is load perpendicular to the middle plane of the plate.

The system of Eq. (16) contains (3+2n) differential equations with the unknown displacements of the middle plane of the plate (u, v, w), and the displacements through the plate thickness U^J and V^J, where J=1, …, n. These equations can be solved by using analytical or numerical and approximate methods.

This chapter shows the analytical solution that is obtained by using the double trigonometric series for simply supported plate, as well as the numerical solution obtained by using the finite element method as one of the currently most applied numerical methods.

3. Analytical solution

3.1. Navier's solution in layerwise theory

Simply supported rectangular laminated plate a x b with N orthotropic layers with angles of orientation θ=0° or 90° is considered. Displacements (Eqs. (1) and (2)) are given in the form of double trigonometric series

$$
\begin{aligned}
u &= \sum_{m,n}^{\infty} X_{mn}\cos\frac{m\pi x}{a}\sin\frac{n\pi y}{b} \\
v &= \sum_{m,n}^{\infty} Y_{mn}\sin\frac{m\pi x}{a}\cos\frac{n\pi y}{b} \\
w &= \sum_{m,n}^{\infty} W_{mn}\sin\frac{m\pi x}{a}\sin\frac{n\pi y}{b} \\
U^{J} &= \sum_{m,n}^{\infty} R_{mn}^{J}\cos\frac{m\pi x}{a}\sin\frac{n\pi y}{b} \\
V^{J} &= \sum_{m,n}^{\infty} S_{mn}^{J}\sin\frac{m\pi x}{a}\cos\frac{n\pi y}{b}
\end{aligned}
\tag{17}
$$

where $\alpha = \frac{m\pi}{a}$; $\beta = \frac{n\pi}{b}$ $J = 1, \ldots, n$

http://dx.doi.org/10.5772/intechopen.69975

Boundary conditions for simply supported plate are met:

$$\begin{aligned} u = w = V^J = N_x = N_x^J = 0 \qquad x = 0, a \qquad J = 1, n \\ v = w = U^J = N_y = N_y^J = 0 \qquad y = 0, b \qquad J = 1, n \end{aligned} \tag{18}$$

Transverse load is displayed in the form of double Fourier series:

$$q(x, y) = \sum_{m,n}^{\infty} Q_{mn} \sin \frac{m\pi x}{a} \sin \frac{n\pi y}{b} \tag{19}$$

where Q_{mn} is the Furie's coefficients which depend on the type of transverse load.

When derivatives of the displacement (Eq. (17)) are substituted into Eq. (16), for each pair (m, n), a system of equations is obtained:

$$\begin{bmatrix} [K] & [K^J] \\ [K^J]^T & [C^{JI}] \end{bmatrix} \begin{Bmatrix} X_{mn} \\ Y_{mn} \\ W_{mn} \\ R_{mn}^J \\ S_{mn}^J \end{Bmatrix} = \begin{Bmatrix} 0 \\ 0 \\ Q_{mn} \\ 0 \\ 0 \end{Bmatrix} \tag{20}$$

Elements of the submatrices K, K^J, and K^{JI} are given in the function of the stiffnesses of laminated plate (Eq. (15)) and variables α and β [26].

By solving Eq. (20), the unknown coefficients X_{mn}, Y_{mn}, W_{mn}, R_{mn}, and S_{mn} for each pair (m, n) are determined. Additionally, the application of Eq. (17) and the summing of the unknown displacements are determined.

The strains are determined using Eq. (6) and then the stresses using Eq. (8). Stresses in the plane of the plate are determined by the expressions:

$$\begin{aligned} \sigma_{xx}(x, y, z) &= -\sum_{m,n}^{\infty} \left\{ \left[\overline{Q}_{11}\alpha \left(X_{mn} + \sum_{J=1}^{n} R_{mn}^J \phi^J(z) \right) + \overline{Q}_{12}\beta \left(Y_{mn} + \sum_{J=1}^{n} S_{mn}^J \phi^J(z) \right) \right] \sin\alpha x \, \sin\beta y \right\} \\ \sigma_{yy}(x, y, z) &= -\sum_{m,n}^{\infty} \left\{ \left[\overline{Q}_{12}\alpha \left(X_{mn} + \sum_{J=1}^{n} R_{mn}^J \phi^J(z) \right) + \overline{Q}_{22}\beta \left(Y_{mn} + \sum_{J=1}^{n} S_{mn}^J \phi^J(z) \right) \right] \sin\alpha x \, \sin\beta y \right\} \\ \sigma(x, y, z) &= -\overline{Q}_{66} \sum_{m,n}^{\infty} \left\{ \left[\beta \left(X_{mn} + \sum_{J=1}^{n} R_{mn}^J \phi^J(z) \right) + \alpha \left(Y_{mn} + \sum_{J=1}^{\infty} S_{mn}^J \phi^J(z) \right) \right] \cos\alpha x \, \sin\beta y \right\} \end{aligned} \tag{21}$$

Shear stresses σ_{xz} and σ_y are constant through the layer J and can be obtained with

$$\begin{aligned} \sigma_{xz}^j(x, y, z) &= \sum_{m,n}^{\infty} \left\{ \overline{Q}_{45}^j \left\{ \frac{1}{h_k} \left[(S_{mn}^{j+1} - S_{mn}^j) + \beta W_{mn} \right] \sin\alpha x \, \cos\beta y \right\} + \overline{Q}_{55}^j \left\{ \frac{1}{h_k} \left[(R_{mn}^{j+1} - R_{mn}^j) + \alpha W_{mn} \right] \cos\alpha x \, \sin\beta y \right\} \right\} \\ \sigma_{xz}^j(x, y, z) &= \sum_{m,n}^{\infty} \left\{ \overline{Q}_{44}^j \left\{ \frac{1}{h_k} \left[(S_{mn}^{j+1} - S_{mn}^j) + \beta W_{mn} \right] \sin\alpha x \, \cos\beta y \right\} + \overline{Q}_{45}^j \left\{ \frac{1}{h_k} \left[(R_{mn}^{j+1} - R_{mn}^j) + \alpha W_{mn} \right] \cos\alpha x \, \sin\beta y \right\} \right\} \end{aligned} \tag{22}$$

A more realistic change of shear stresses σ_{xz} and σ_y through the plate thickness can be obtained using approximate procedures. In this chapter, approximate procedure as described in Ref. [23] was used.

3.2. Numerical examples

For a simply supported rectangular plate, stresses are determined in dimensionless form:

$$\overline{\sigma}_{xx} = \frac{1}{qs^2}\sigma_{xx}, \qquad \overline{\sigma}_{yy} = \frac{1}{qs^2}\sigma_{yy}, \qquad \overline{\sigma}_{xz} = \frac{1}{qs}\sigma_{xz}, \qquad \overline{\sigma}_{yz} = \frac{1}{qs}\sigma_{yz} \tag{23}$$

where $s=a/h$, $a=b$ are plate dimensions in the plane of the plate; h is the plate thickness

All layers of the laminated plate have the same thickness and material properties. For predefined material characteristics of layers E_1, E_2, G_{12}, G_{21}, G_{23}, ν_{12}, and ν_{13}, changes of dimensionless stresses σ_{xx}, σ_{yy}, σ_{yy}, and σ_{xz} are displayed in the points with coordinates $A=1{,}105662(a/2)$ and $B=1{,}894338(a/2)$

$$\overline{\sigma}_{xx} = \overline{\sigma}_{xx}(A, A, z), \quad \overline{\sigma}_{yy} = \overline{\sigma}_{yy}(A, A, z), \quad \overline{\sigma}_{xz} = \overline{\sigma}_{xz}(B, A, z), \quad \overline{\sigma}_{yz} = \overline{\sigma}_{yz}(A, B, z) \tag{24}$$

Example 1: The square laminated composite plate with five layers 0°/90°/0°/90°/0° and the aspect ratio $a/h=4$ loaded by a bi-sinusoidal is considered. Characteristics of the layers are $E_1/E_2=25$, $E_2=1$, $G_{12}=G_{13}=0.5$, $G_{23}=0.2$, $\nu_{12}=\nu_{13}=0.25$. The change of dimensionless normal stresses σ_{xx}, σ_{yy} and shear stresses σ_{yz}, σ_{xz} are displayed in **Figures 2** and **3**.

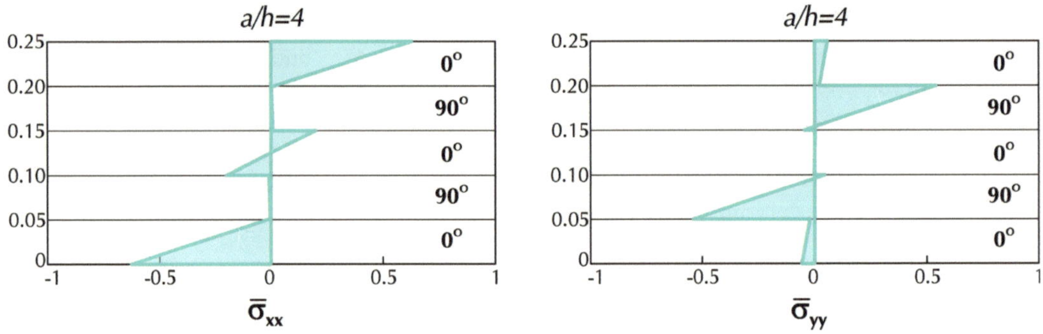

Figure 2. $\overline{\sigma}_{xx}$ and $\overline{\sigma}_{yy}$ through the thickness of the plate 0°/90°/0°/90°/0°, bi-sinusoidal load.

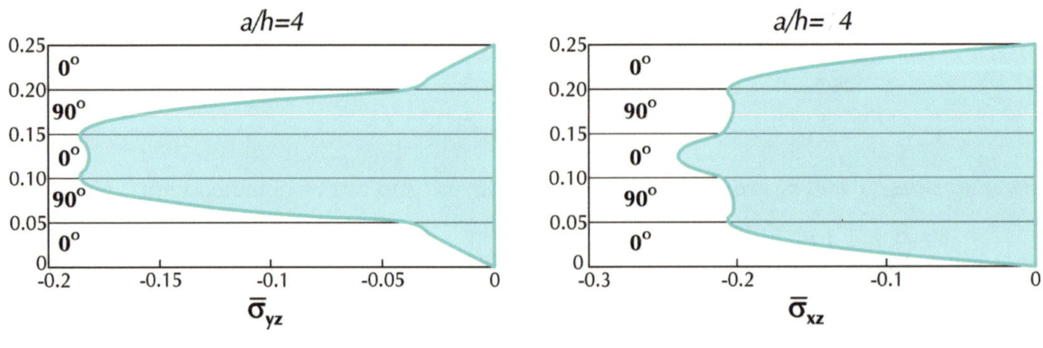

Figure 3. $\overline{\sigma}_{yz}$ and $\overline{\sigma}_{xz}$ through the thickness of the plate 0°/90°/0°/90°/0°, bi-sinusoidal load.

Example 2: The square laminated composite plate with six layers 0°/90°/0°/90°/0°/90° and the aspect ratio a/h=10 loaded by uniformly distributed load of intensity q is considered. The material properties of the layers are as in Example 1. **Figures 4** and **5** graphically show the change of dimensionless stresses σ_{xx}, σ_{yy}, σ_{yz}, and σ_{xz} through the thickness of the plate.

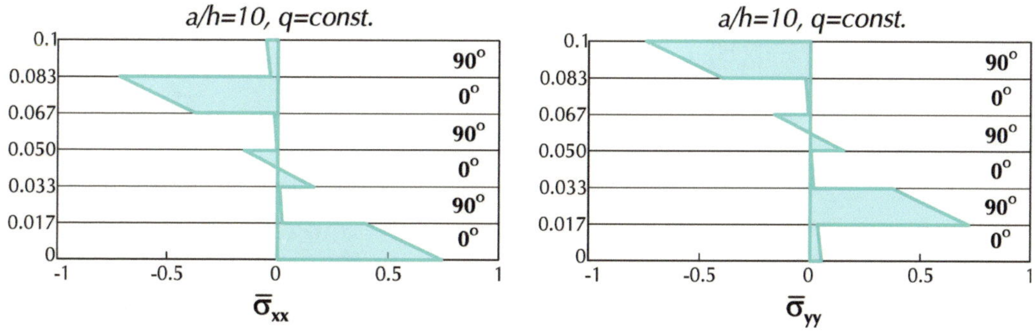

Figure 4. $\overline{\sigma}_{xx}$ and $\overline{\sigma}_{yy}$ through the thickness of the plate 0°/90°/0°/90°/0°/90°.

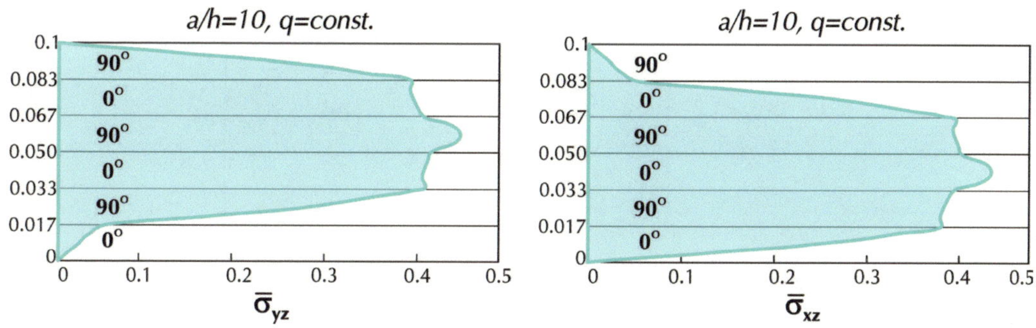

Figure 5. $\overline{\sigma}_{yz}$ and $\overline{\sigma}_{xz}$ through the thickness of the plate 0°/90°/0°/90°/0°/90°.

4. Finite element method

4.1. Stiffness matrix of the kinematic variable finite element

The components of displacement of finite element can be written as a combination of two-dimensional interpolation function ϕ_i and nodal displacements u_i, v_i, w_i, $u_i{}^j$, $v_i{}^j$

$$\left(u, v, w, u_j, v_j\right) = \sum_{i=1}^{m} \left(u_i, v_i, w_i, u_i^j, v_i^j\right)\phi_i \tag{25}$$

Interpolation functions ϕ_i are defined in the natural coordinates (ξ,η). The isoparametric formulation of the finite element is adopted, so that the geometry is interpolated the same way as the primary variables (displacements, see Eq. (25)):

$$x = \sum_{i=1}^{m} x_i \phi_i(\xi, \eta)$$
$$y = \sum_{i=1}^{m} y_i \phi_i(\xi, \eta) \quad (26)$$

where x, y are point coordinates and x_i, y_i are coordinates of the finite element.

According to Eqs. (6) and (25), matrices H and $\overline{H}$ are defined as follows:

$$[H]_{(5x3m)} = \begin{bmatrix} \frac{\partial\phi_1}{\partial x} & 0 & 0 & \frac{\partial\phi_2}{\partial x} & 0 & 0 & \cdots & \frac{\partial\phi_m}{\partial x} & 0 & 0 \\ 0 & \frac{\partial\phi_1}{\partial y} & 0 & 0 & \frac{\partial\phi_2}{\partial y} & 0 & \cdots & 0 & \frac{\partial\phi_m}{\partial y} & 0 \\ \frac{\partial\phi_1}{\partial y} & \frac{\partial\phi_1}{\partial x} & 0 & \frac{\partial\phi_2}{\partial y} & \frac{\partial\phi_2}{\partial x} & 0 & \cdots & \frac{\partial\phi_m}{\partial y} & \frac{\partial\phi_m}{\partial x} & 0 \\ 0 & 0 & \frac{\partial\phi_1}{\partial x} & 0 & 0 & \frac{\partial\phi_2}{\partial x} & \cdots & 0 & 0 & \frac{\partial\phi_m}{\partial x} \\ 0 & 0 & \frac{\partial\phi_1}{\partial y} & 0 & 0 & \frac{\partial\phi_2}{\partial y} & \cdots & 0 & 0 & \frac{\partial\phi_m}{\partial y} \end{bmatrix} \quad (27)$$

$$[\overline{H}]_{(5x2m)} = \begin{bmatrix} \frac{\partial\phi_1}{\partial x} & 0 & \frac{\partial\phi_2}{\partial x} & 0 & \cdots & \frac{\partial\phi_m}{\partial x} & 0 \\ 0 & \frac{\partial\phi_1}{\partial y} & 0 & \frac{\partial\phi_2}{\partial y} & \cdots & 0 & \frac{\partial\phi_m}{\partial y} \\ \frac{\partial\phi_1}{\partial y} & \frac{\partial\phi_1}{\partial x} & \frac{\partial\phi_2}{\partial y} & \frac{\partial\phi_2}{\partial x} & \cdots & \frac{\partial\phi_m}{\partial y} & \frac{\partial\phi_m}{\partial x} \\ \phi_1 & 0 & \phi_2 & 0 & \cdots & \phi_m & 0 \\ 0 & \phi_1 & 0 & \phi_2 & \cdots & 0 & \phi_m \end{bmatrix} \quad (28)$$

where m is the number of nodes in the plane of the plate.

The first derivatives of the ϕ_i with respect to the global coordinates x, y can be obtained by using the local coordinates ξ, η as follows:

$$\begin{Bmatrix} \frac{\partial\phi_i}{\partial x} \\ \frac{\partial\phi_i}{\partial y} \end{Bmatrix} = \begin{bmatrix} x,_\xi & y,_\xi \\ x,_\eta & y,_\eta \end{bmatrix}^{-1} \begin{Bmatrix} \phi,^i_\xi \\ \phi,^i_\eta \end{Bmatrix} = [J]^{-1} \begin{Bmatrix} \phi,^i_\xi \\ \phi,^i_\eta \end{Bmatrix} \quad (29)$$

where J is the Jacobian matrix of the transformation in the natural coordinates:

$$[J] = \begin{bmatrix} \frac{\partial x}{\partial \xi} & \frac{\partial y}{\partial \xi} \\ \frac{\partial x}{\partial \eta} & \frac{\partial y}{\partial \eta} \end{bmatrix} \quad (30)$$

The second derivatives of the interpolation functions, which are used during determinations of the shear stresses perpendicular to the plane plate, are

$$\begin{Bmatrix} \phi,_{xx} \\ \phi,_{yy} \\ \phi,_{xy} \end{Bmatrix} = [J_1]^{-1} \left(\begin{Bmatrix} \phi_{i'},_{\xi\xi} \\ \phi_{i'},_{\eta\eta} \\ \phi_{i'},_{\xi\eta} \end{Bmatrix} - [J_2] \begin{Bmatrix} \phi_{i'},_{x} \\ \phi_{i'},_{y} \end{Bmatrix} \right) \tag{31}$$

where

$$[J_1] = \begin{bmatrix} x,_{\xi}^2 & y,_{\xi}^2 & 2y,_{\xi}^2 x,_{\xi}^2 \\ x,_{\eta}^2 & y,_{\eta}^2 & 2y,_{\eta}^2 x,_{\eta}^2 \\ x,_{\eta\xi} & y,_{\eta\xi} & x,_{\eta\xi} + y,_{\eta\xi} \end{bmatrix} \tag{32}$$

$$[J_2] = \begin{bmatrix} x,_{\xi\xi} & y,_{\xi\xi} \\ x,_{\eta\eta} & y,_{\eta\eta} \\ x,_{\eta\xi} & y,_{\eta\xi} \end{bmatrix} \tag{33}$$

According to the principle of stationary potential energy, the first variation of the potential energy must be zero for equilibrium conditions. The first variation of the potential energy of the finite element is

$$\delta \Pi_e = \int_{V_e} \varepsilon^T \overline{Q} \delta\varepsilon \, dV \tag{34}$$

By using Eq. (7), the strains can be written in the following form:

$$\{\varepsilon\} = [H]\{\Delta\} + [\overline{H}] \sum_{j=1}^{n} \phi^J \{\Delta^J\} \tag{35}$$

where ϕ^J are the interpolation functions perpendicular to the plane of the plate or its derivatives:

$$\phi^J(z) = \begin{bmatrix} \psi^j(z) \\ \psi^j(z),_z \end{bmatrix} \tag{36}$$

Inserting Eq. (35) into Eq. (34), the first variation of the potential energy of the finite element becomes

$$\delta \Pi_e = \int_{V_e} \left(\Delta^T H^T \overline{H}^T \sum_{J=1}^{n} \phi^J \Delta^{JT} \right) \cdot \overline{Q} \cdot \left(H\delta\Delta + \overline{H} \sum_{I=1}^{n} \phi^I \delta\Delta^I \right) dV \tag{37}$$

After multiplication, the stiffness matrix of the finite element takes the form:

$$\begin{aligned} k_e = \iiint_{V_e} \left(H^T \overline{Q} H + \sum_{J=1}^{n} H^T \overline{Q} \phi^J \overline{H} + \sum_{I=1}^{n} \overline{H}^T \overline{Q} \phi^I H + \sum_{I=1}^{n} \sum_{J=1}^{n} \overline{H}^T \overline{Q} \phi^I \phi^J \overline{H} \right) dV = \\ \iint_{A_e} \left(H^T A H + \sum_{J=1}^{n} H^T B^J \overline{H} + \sum_{I=1}^{n} \overline{H}^T B^I H + \sum_{I=1}^{n} \sum_{J=1}^{n} \overline{H}^T D^{IJ} \overline{H} \right) dA \end{aligned} \tag{38}$$

The stiffness matrix of the finite element can be written in a matrix from as

$$k_e = \begin{bmatrix} k^{11} & k_1^{12} & k_2^{12} & \dots & k_n^{12} \\ k_1^{21} & k_{11}^{22} & k_{12}^{22} & \dots & k_{1n}^{22} \\ k_2^{21} & k_{12}^{22} & k_{22}^{22} & \dots & k_{2n}^{22} \\ \dots & \dots & \dots & \dots & \dots \\ k_n^{21} & k_{n1}^{22} & k_{n2}^{22} & \dots & k_{nn}^{22} \end{bmatrix} \tag{39}$$

From the structure of the matrix (Eq. (39)), four submatrices can be noticed, where elements (see Eq. (38)) are determined as follows:

$$\begin{aligned} k^{11} &= \iint_{A_e} H^T A H dA \qquad i = 1, \dots, n \\ k_i^{12} &= \iint_{A_e} H^T B^i \overline{H} dA \qquad i = 1, \dots, n \\ k_i^{21} &= \iint_{A_e} \overline{H}^T B^i H dA \qquad i = 1, \dots, n \\ k_{ij}^{22} &= \iint_{A_e} \overline{H}^T D^{ij} \overline{H} dA \qquad i, j = 1, \dots, n \end{aligned} \tag{40}$$

The area integral can be written in the form of the natural coordinates:

$$dA = dx\,dy = \det J\, d\xi\, d\eta \tag{41}$$

as follows:

$$\iint_{A_e} (\dots) dA = \int_{=1}^{+1} \int_{-1}^{+1} (\dots) \det J\, d\xi\, d\eta \tag{42}$$

where *det J* is the determinant of the Jacobian matrix of the transformation according to Eqs. (26) and (30):

$$|J| = \begin{vmatrix} \sum_{i=1}^{m} x_i \frac{\partial \phi}{\partial \xi} & \sum_{i=1}^{m} y_i \frac{\partial \phi}{\partial \xi} \\ \sum_{i=1}^{m} x_i \frac{\partial \phi}{\partial \eta} & \sum_{i=1}^{m} y_i \frac{\partial \phi}{\partial \eta} \end{vmatrix} \tag{43}$$

Thereafter, the submatrices of the stiffness matrix (Eq. (39)) are given by

$$
\begin{aligned}
k^{11} &= \int_{-1}^{+1}\int_{-1}^{+1} H^T A H \det J \, d\xi \, d\eta \qquad i = 1, \dots, n \\
k_i^{12} &= \int_{-1}^{+1}\int_{-1}^{+1} H^T B^i \overline{H} \det J \, d\xi \, d\eta \qquad i = 1, \dots, n \\
k_i^{21} &= \int_{-1}^{+1}\int_{-1}^{+1} \overline{H}^T B^{iT} H \det J \, d\xi \, d\eta \qquad i = 1, \dots, n \\
k_{ij}^{22} &= \int_{-1}^{+1}\int_{-1}^{+1} \overline{H}^T D^{ij} \overline{H} \det J \, d\xi \, d\eta \qquad i, j = 1, \dots, n
\end{aligned}
\tag{44}
$$

4.2. Interpolation functions and numerical integration

Interpolation or shape functions in finite element analysis depend on the dimensionality of the problem and type of elements used for discretization of the problem. Layerwise theory introduces two types of interpolation functions. The first type is the interpolation functions in the plane (x,y) that can be selected from the family of well-known two-dimensional shape functions. The degree and type of interpolation functions depend on the type of finite element and the number of nodes in which variables are defined. The second type is the interpolation functions through the plate thickness. These interpolation functions are the functions of one variable (z) and can be linear, quadratic, or higher order functions.

For the models of rectangular finite element, it is possible to implement a wide range of standard 2D interpolation functions in the plane of the plate, as well as 1D interpolation functions through the thickness of the layer. At our disposal are 2D finite elements such as E4—four-node Lagrange rectangular element, E8—eight-node Serendipity rectangular element, E9—nine-node Lagrange rectangular element, E12—12-node Serendipity rectangular element, and E16—16-node Lagrange rectangular element. Each of these elements can be combined with one or more 1D Lagrange elements: L—linear element, Q—square element, C—cubic element, in order to create a wide spectrum of a variety of finite elements based on the layerwise theory. We use notation E16-L4 to denote 2D element with 16 nodes combined with four linear 1D elements through the thickness of the plate.

For laminated plate containing N layers with adopted linear interpolation through the thickness (number of nodes is $n=N+1$) and the interpolation with m nodes in the plane of the plate, the total number of degrees of freedom of one finite element is $(3+2n)m$. For the finite element displayed in **Figure 6**, $m=8$ and $N=3$, which leads to the number of degrees of freedom $(3+2\times 4)\times 8=88$. It is clear that the number of degrees of freedom of the finite element is rapidly growing with the increasing degree of interpolation functions.

To determine the stiffness matrix of elements, it is necessary to perform numerical integration. If Gauss-Legendre integration is used, then the integral is calculated by

$$I(\xi,\eta)=\int_{-1}^{+1}\int_{-1}^{+1}f(\xi,\eta)d\xi d\eta=\sum_{i=1}^{m}\sum_{j=1}^{m}\omega_i\omega_j f(\overline{\xi},\overline{\eta}) \qquad (45)$$

where ω_i and ω_j are the weights of the integration, and $(\overline{\xi},\overline{\eta})$ coordinates of the points at which the integration is performed. The weights and the position of integration points can be found in the wide literature for the finite element method.

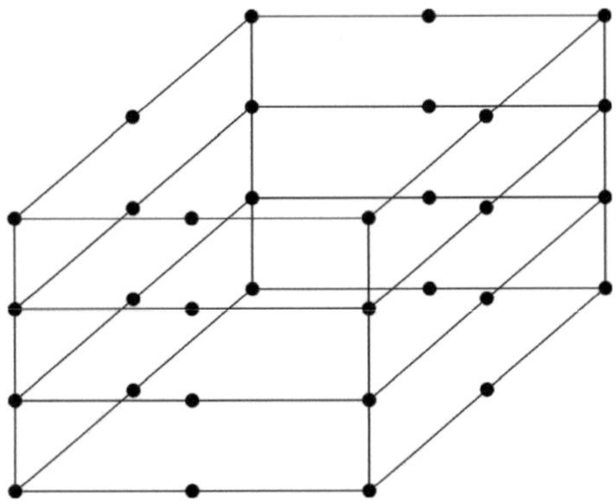

Figure 6. Rectangular finite element E8–L3.

4.3. Governing equations

Governing equations in the finite element method are calculated from the principle of virtual displacements. The first variation of the potential of the system of finite elements is

$$\partial\Pi=\delta \boldsymbol{U}^T\boldsymbol{K}\boldsymbol{U}-\delta \boldsymbol{U}^T\boldsymbol{F} \qquad (46)$$

By applying the principle of virtual displacements, the system of algebraic equations of the finite element method is obtained:

$$\boldsymbol{K}\boldsymbol{U}=\boldsymbol{F} \qquad (47)$$

where $\boldsymbol{U}$ is the vector of the nodal displacements, $\boldsymbol{F}$ is the vector of the nodal forces, and $\boldsymbol{K}$ is the stiffness matrix, all for system of assembly of finite elements formed in the global coordinate system.

Methods of forming the stiffness matrix $\boldsymbol{K}$ and the vectors $\boldsymbol{U}$ and $\boldsymbol{F}$ are generally known and can be found in the expanded literature that deal with the finite element method. Solutions of the system of algebraic equations (Eq. (46)) are the nodal displacements.

4.4. Numerical examples

A square four-layer simply supported laminated plate with the arrangement of layers 60°/-45°/-45°/60° with boundary conditions $u = v = w = 0$ for $x = 0, a$ and $y = 0, b$, loaded by the uniformly distributed load intensity q, was considered. The adopted finite element is E4–L1 with layers of the same thickness and the same material properties. Due to the demanding and

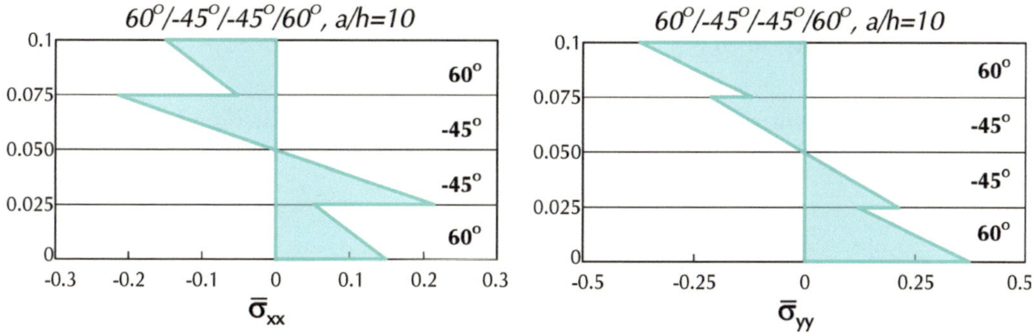

Figure 7. $\overline{\sigma}_{xx}$ and $\overline{\sigma}_{yy}$ through the thickness of the plate.

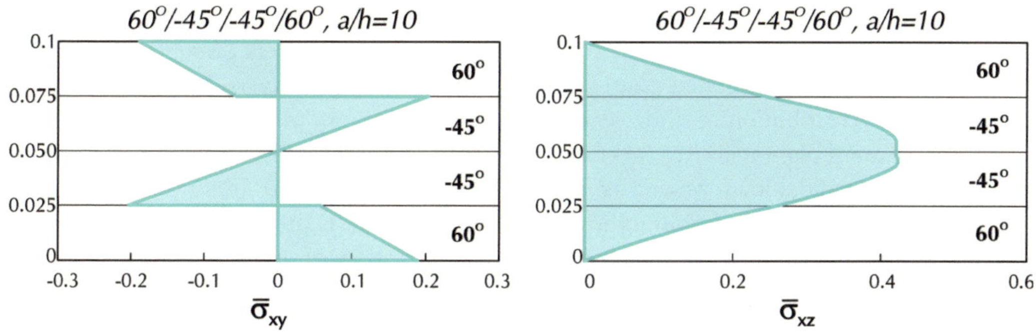

Figure 8. $\overline{\sigma}_{xy}$ and $\overline{\sigma}_{xz}$ through the thickness of the plate.

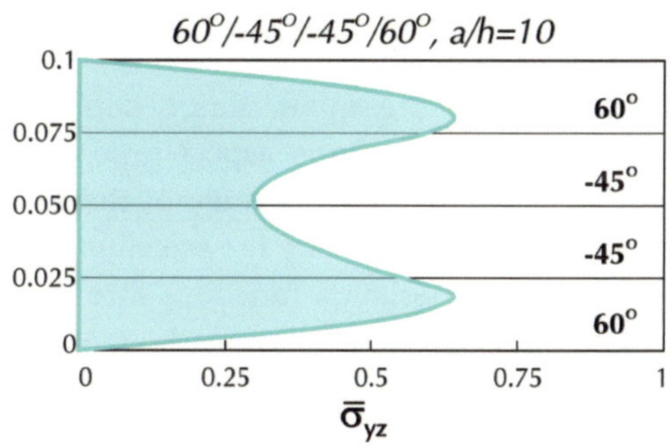

Figure 9. $\overline{\sigma}_{yz}$ through the thickness of the plate.

very extensive process of calculation with a large number of unknowns, a relatively rough mesh of finite elements with 4×4 finite elements is adopted.

Material characteristics of the layers are E_1=25, E_2=1, G_{12}=G_{13}=0.5, G_{23}=0.2, ν_{12}=ν_{13}=0.25.

Stresses are determined in a dimensionless form according to Eq. (23) described in Section 2 of this chapter, wherein the shear stress in the plane of the plate is also given in the same dimensionless form as the normal stresses. Stresses are calculated in points with coordinates:

$\overline{\sigma}_{xx}(0.4471688a; 0.4471688a)$ $\overline{\sigma}_{yy}(0.4471688a; 0.4471688a)$ $\overline{\sigma}_{xy}(0.05283122a; 0.05283122a)$
$\overline{\sigma}_{yz}(0.4471688a; 0.9471688a)$ $\overline{\sigma}_{xz}(0.9471688a; 0.4471688a)$

Changing of dimensionless stresses through the thickness of the laminated plate is presented in **Figures 7–9**.

5. Conclusion

This chapter refers to the application of Reddy's layerwise theory, based on the assumption that dilatation perpendicular to the middle plane of the laminated plate is neglected. In the bending equations for laminated plates, the primary variables of node J are displacements through the thickness of the plate $u^J(x,y,z)$ and $v^J(x,y,z)$ and deflections $w^J(x,y,z)=w(x,y)$.

Analytical solution of the bending equations in the layerwise theory is presented for simply supported rectangular laminated plate which contains orthogonal layers. A graphical representation of the distribution of dimensionless stresses σ_{xx}, σ_{yy}, σ_{yz}, and σ_{xz} through the thickness for the square plate with five layers 0°/90°/0°/90°/0° and the aspect ratio $a/h=4$, which is loaded by bi-sinusoidal load, as well as for six-layer plate 0°/90°/0°/90°/0°/90° with the aspect ratio $a/h=10$, which is loaded with uniformly distributed load, is presented. Obtained results indicate that the analytical solution provides determination of more realistic change of interlayer stresses, and it also allows the incorporation of delamination in the calculation procedures. However, the analytical solution is difficult to define for complex geometry of the plates, for arbitrary angle of orientation layers, for arbitrary boundary conditions, as well as for various forms of nonlinearity. These are the reasons why in layerwise theory the use of numerical method such as finite element method is usually recommended.

In Chapter 4, a graphical representation of the distribution of dimensionless stresses σ_{xx}, σ_{yy}, σ_{xy}, σ_{yz}, and σ_{xz} through the thickness of the square, simply supported rectangular laminated plate with four layers 60°/-45°/-45°/60° and the aspect ratio $a/h=10$, which is loaded by uniformly distributed load, is presented. The accuracy of the solution, obtained by using the finite element method, is highly influenced by the selection of interpolation functions in the plane (2D), by the selection of interpolation functions through the thickness (1D) of the plate. Additionally, as in general for this calculation method, a number of finite elements of the analyzed domain highly influence the accuracy of obtained solution. However, arbitrary increase of the number of finite element nodes in the plane of the plate and through the thickness of the plate suddenly drastically increases the number of degrees of freedom, which gives rise to the uneconomical nature of the finite element method for more complex mechanical and engineering problems. Regardless of this deficiency, the finite element method is a more general and acceptable for the analysis of laminated composite plates than the proposed analytical method.

Aside the drawbacks and disadvantages of the analytical solution, the same should not be excluded from the engineering and scientific practice, since it represents the ultimate convergence and accuracy criteria for any other numerical or approximate methods, including the finite element method presented in this chapter.

http://dx.doi.org/10.5772/intechopen.69975

Author details

Marina Rakočević

Address all correspondence to: marinara@ac.me

Faculty of Civil Engineering, University of Montenegro, Podgorica, Montenegro

References

[1] Reddy JN. Mechanics of Laminated Composite Plate: Theory and Analysis. New York, NY: CRC Press; 1997. ISBN 0-8493-3101-3

[2] Whitney JM. The effect of transverse shear deformation in the bending of laminates plates. Journal of Composite Materials. 2004;**3**:534-547

[3] Reissner E. A consistent treatment of transverse shear deformations in laminated anisotropic plates. The American Institute of Aeronautics and Astronautics Journal. 1972;**10**(5): 716-718

[4] Reddy JN. Energy and Variational Methods in Applied Mechanics. New York, NY: John Wiley; 1984

[5] Christensen RM, Lo KH, Wu EM. A high-order theory of plate deformation, part1: Homogeneous plates. Journal of Applied Mechanics. 1977;**44**(7):636-668

[6] Christensen RM, Lo KH, Wu EML. A high-order theory of plate deformation, part 2: Laminated plates. Journal of Applied Mechanics. 1977;**44**(4):669-676

[7] Vuksanović DJ. Linear analysis of laminated composite plates using single layer higher-order discrete models. Composite Structures. 2000;**48**:205-211

[8] Reddy JN. A simple higher-order theory for laminated composite plates. Journal of Applied Mechanics. 1984;**51**:745-752

[9] Reddy JN. A refined nonlinear theory of plates with transverse shear deformation. International Journal of Solids and Structure. 1984;**20**(9):881-906

[10] Reddy JN, Barbero EJ, Teply JL. A plate bending element based on a generalized laminated plate theory. International Journal for Numerical Methods in Engineering. 1989;**28**:2275-2292

[11] Reddy JN, Barbero EJ, Teply JL. An accurate determination of stresses in thick laminates using a generalized plate theory. International Journal for Numerical Methods in Engineering. 1990;**29**:1-14

[12] Redyy JN, Barbero EJ. Modelling of thick composites using a layerwise laminate theory. International Journal for Numerical Methods in Engineering. 1993;**36**:655-677

[13] Reddy JN, Robbins DH. Theories and computational models for composite laminates. Applied Mechanics Review. 1994;**47**(6):147-165

[14] Reddy JN. Theory and analysis of laminated composite plates. Mechanics of Composite Materials and Structures. 1999;**361**:1-79

[15] Ferreira AJM, Fasshauer GE, Batra RC, Rodrigues JD. Static deformations and vibration analysis of composite and sandwich plates using a layerwise theory and RBF-PS discretizations with optimal shape parameter. Composite Structures. 2008;**86**:328-343

[16] Vuksanović Đ, Rakočević M. Kompozitne ploče sa delaminacijama-analitičko rješenje za slobodno oslonjene ploče. In: Simpozijum o istraživanjima i primjeni savremenih dostignuća u našem građevinarstvu u oblasti materijala i konstrukcija (JUDIMK); Oktobar 2002. pp. 131-136

[17] Vuksanović Đ, Rakočević M. Opšta teorija laminarnih ploča–analitičko rješenje za slobodno oslonjene ploće, In: Simpozijum o istraživanjima i primjeni savremenih dostignuća u našem graevinarstvu u oblasti materijala i konstrukcija, JUDIMK; Oktobar 2002. pp. 125-130

[18] Vuksanović Đ, Ćetković M. Analytical solution for multilayer plates using general layerwise plate theory. Facta Universitates Series. Architecture and Civil Engineering. 2005;**3**(2):121-136

[19] Cetković M, Vuksanović D. Bending, free vibrations and buckling of laminated composite and sandwich plates using a layerwise displacement model. Composite Structures. 2009;**88**(2):219-227

[20] Cetković M, Vuksanović D. Large deflection analysis of laminated composite plates using layerwise displacement model. Structural Engineering & Mechanics. 2011;**40**(2):257-277

[21] Rakočević M. Stress in multi-layered composite plates. Gradevinar. 2005;**57**(7):503-509

[22] Rakočević M. Analysis of laminated composite plate. Gradevinar. 2011;**63**(9/10):819-825

[23] Rakocevic M. Approximate procedure for calculation of shear stresses σ_{xz} and σ_{yz}. Journal of Applied Engineering Science. 2012;**10**(1):37-42

[24] Rakočević M. Analysis of simply supported laminated composite plates using double trigonometric series. In: Proceedings of the 16th International Symposium of Macedonian Association of Structural Engineers (MASE 2016); October 2015; Ohrid, Macedonia

[25] Rakocevic M. Analytical solution for simply supported laminated composite plate based on partial layerwise theory. Journal of Applied Engineering Science. 2016;**14**(1):102-108

[26] Rakočević M, Popović S, Ivanišević N. A computational method for laminated composite plates based on layerwise theory. Composites Part B. 2017;**122**:202-218

Section 3

Application of Lamination

Chapter 4

Dynamic Modeling of a Serial Link Robot Laminated with Plastic Film

Norimitsu Sakagami and Mizuho Shibata

Additional information is available at the end of the chapter

http://dx.doi.org/10.5772/intechopen.72441

Abstract

This chapter presents serial link robots laminated with a plastic film, a derivation of the equations of motion of the laminated robots, and numerical simulation. Recently, to become capable of wide application for several serial link robots that work outside, waterproofing and dustproofing techniques are required. We have proposed a robot packaging method to improve waterproof and dustproof properties of serial link robots. Using the proposed packaging method, rigid links with some active joints are loosely laminated with plastic film to protect the links from dust and water. In the next step of our research, we must derive the equations of motion of the laminated robots for the design and performance improvement from the viewpoint of high speed and high energy efficiency. We assume a plastic film as a closed-loop link structure with passive joints in this chapter. A rigid serial link (fin) connected with a motor-actuated joint moves a closed-loop link structure with passive joints. We numerically investigate the influence of the flexural rigidity of a plastic film on the motion of the rigid fin. This research not only contributes to the lamination techniques but also develops a novel application of waterproofing and dustproofing techniques in robotics.

Keywords: closed-loop mechanism, equations of motion, serial link robot, flexible mechanism, vacuum packaging, fish-like robots

1. Introduction

This chapter presents a description of equations of motion of serial link robots laminated with plastic film. Serial link robots are designed as a series of links connected by motor-actuated joints. Typical applications of serial link robots are serial manipulators, which are the most common industrial robots, such as pick-and-place assembly robots [1, 2] and welding robots [3, 4]. A salient feature of serial link robots is their large workspace in comparison with the robot size [5]. Additionally, serial link robots consist of simple structures. For that reason, they have also been used as various applications as humanoids [6, 7], robotic hands [8, 9], biped robots [10, 11], robotic legs [12, 13], snake robots [14, 15], fish-like robots [16, 17], and a jumping robot [18]. Kinematics of a serial link robot itself has remained a hot topic in robotics and has been studied for the last few decades [19].

Recently, to become capable of wide applications for several serial link robots that must be used outdoors, waterproofing and dustproofing techniques must be improved. These techniques have been developed for ocean engineering and field engineering for the past few decades. For example, most underwater robots have waterproof and pressure-tight housings made of metal such as stainless steel or titanium alloy [20–22]. In another waterproofing method used in ocean engineering, called the pressure equalization method, waterproof housings are filled with an insulating fluid such as an industrial oil or a cleaning fluid used for semiconductors. This method has been applied to underwater equipment of several types such as undersea batteries of a submarine [23] and light devices.

We have proposed a robot packaging method to improve waterproofing of a serial link robot [24]. In the robot packaging method, a serial link mechanism connected with motor-actuated joints is packaged in plastic film in a chamber of a vacuum packaging machine, so that the serial link robot is laminated with a plastic film (**Figure 1**). We applied the packaging method to fabrication of a fish-like robot (**Figure 2(a)**). The body (the float in **Figure 2(b)**) and fin (the oscillation plate in **Figure 2(b)**) are connected via a servo motor in series. The outer plastic film is inflected by the motion of the inside fin. The fish-like robot generates thrust using the body inflection underwater. This packaging method is applicable not only to a rotary joint

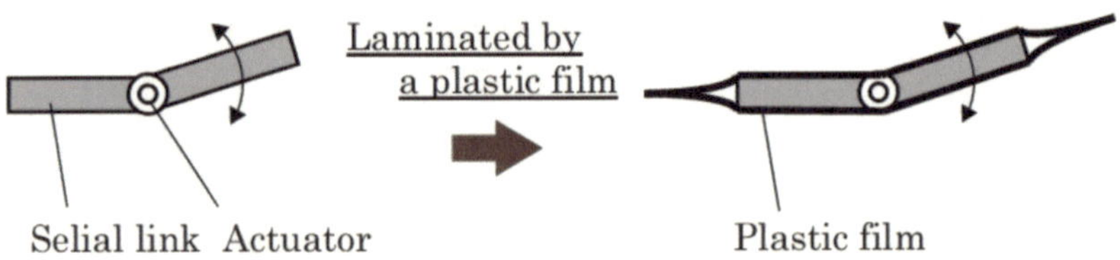

Figure 1. Concept of a serial link robot laminated with plastic film.

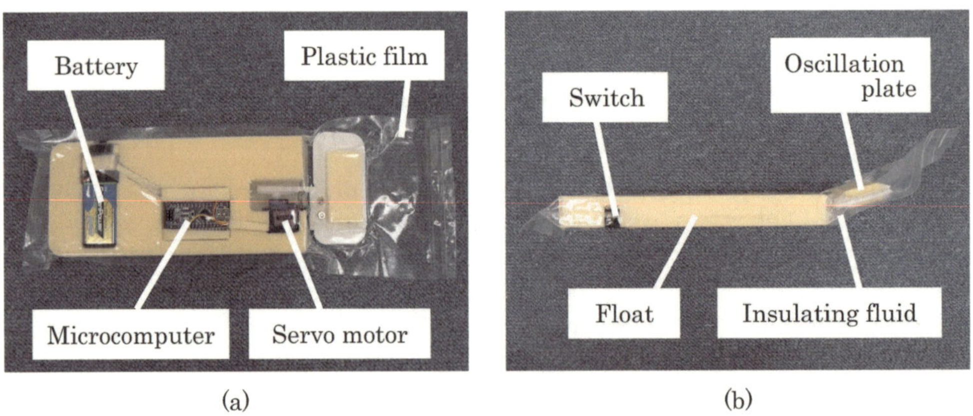

Figure 2. Concept of a fish-like robot laminated with plastic film.

mechanism but also to a prismatic joint mechanism. A ball-screw mechanism was packaged in plastic film to fabricate an attitude control system for underwater robots [25].

In the robot fabricated using the packaging method, the rigid links and the plastic film are loosely laminated. The serial link in the plastic film is packaged in the chamber of the vacuum packaging machine after the decompression process. No adhesion exists between the rigid link and the plastic film. However, the rigid link and the plastic film contact mutually with little slippage between them after depressurization during the robot packaging method. To move the serial link with an actuator in the plastic film, the actuator torque must overcome the static frictional force between the links and the film. We encapsulate an insulating fluid to improve lubricity between the links and the film. Therefore, we must consider not only the material properties of the plastic film but also the effects of the insulating fluid to improve the performance of the serial link robots.

In the next step of our research, we must derive equations of motion to achieve performance improvements such as thrust force and energy efficiencies of the laminated robot system. Over the past few decades, several researchers have studied models of laminated structures. In magnetics, modeling of hysteresis losses [26–28], eddy current losses [29, 30], and temperature effects [31] for magnetic laminations have been proposed under some conditions. Depending on those conditions, a 1-D model [32], a 2-D finite-element model [33], and a 3-D finite-element model [34] have been selected to analyze the laminated structure performance. However, we must derive the equations of motion and analyze the motions of laminated structures including a serial link robot from the viewpoint of robotics. In this chapter, we propose equations of motion of the laminated robot underwater. In addition, based on the equations of motion, we numerically estimate the influence of flexural rigidity of plastic films for the motion performance of the serial link robot.

This research not only contributes to lamination techniques but also develops a novel application of waterproofing and dustproofing techniques for application to robotics. Several techniques using an elastic material such as silicone have been applied to robots for waterproofing [35], impact absorption [36, 37], and decoration [38]. In terms of most of the robots that are covered with an elastic material, the actuator force and torque must overcome the elastic force during motions of the robot body. Thin plastic films can be flexible, but they have lower elasticity for bending than other elastic materials such as silicone. Therefore, we can select a low-force and torque actuator to inflect the robot body laminated with a thin plastic film.

This chapter is organized as follows: Section 2 briefly outlines the concept of a serial link robot laminated with a plastic film. This plastic film has flexible and nonextendable properties, which are useful for vacuum packaging in food industry. Also Section 2 takes applications of two serial link robots, such as fish-like robots we have proposed. Section 3 discusses derivation of the equations of motion of the serial link robots including a plastic film that laminates the robot body. Here, we assume that the plastic film has a closed-loop link structure with passive joints. A serial link as a fin connected by a motor-actuated joint drives the closed-loop link structure (the plastic fin) with passive joints. We consider equations of motion to estimate the motion characteristics of the proposed robot system. Section 4 presents validation of the equations of motion through several simulations. Section 5 presents a summary our future work and conclusions.

2. Application

In this section, we briefly describe the concept of lamination of a serial link robot for waterproofing and pressure tightness. In robotics, piezoelectric actuators have been used to drive several robots as a useful application of lamination techniques [39, 40]. We used a vacuum packaging machine to laminate a serial link robot with a plastic film. We designated this fabrication as "robot packaging."

2.1. Robot packaging method

Figure 3 presents the "robot packaging" process as an example of fabricating a fish-like robot. The entire robot body was covered by a flexible plastic film. The process is divisible into four steps: (a) encapsulation of the internal components, including a microcontroller, a drive circuit, a battery, a servomotor, and an oscillation plate, in a plastic film bag; (b) pouring of an insulating fluid, specifically industrial oil or cleaning fluid for semiconductors, into the plastic bag; (c) depressurization of the inside of the robot using a vacuum packaging machine. This process reduces the quantity of air in the film bag; and (d) sealing of the plastic film by a sealer within the chamber of the vacuum packaging machine after depressurizing.

This plastic film has flexible and nonextendable properties, which is used for vacuum packaging in food industry. This packaging method corresponds with the method used in the food industry [41]. The internal components including electrical circuits in the body of the robot are not shortened by the insulating fluid surrounding the circuits. Using this method, we can

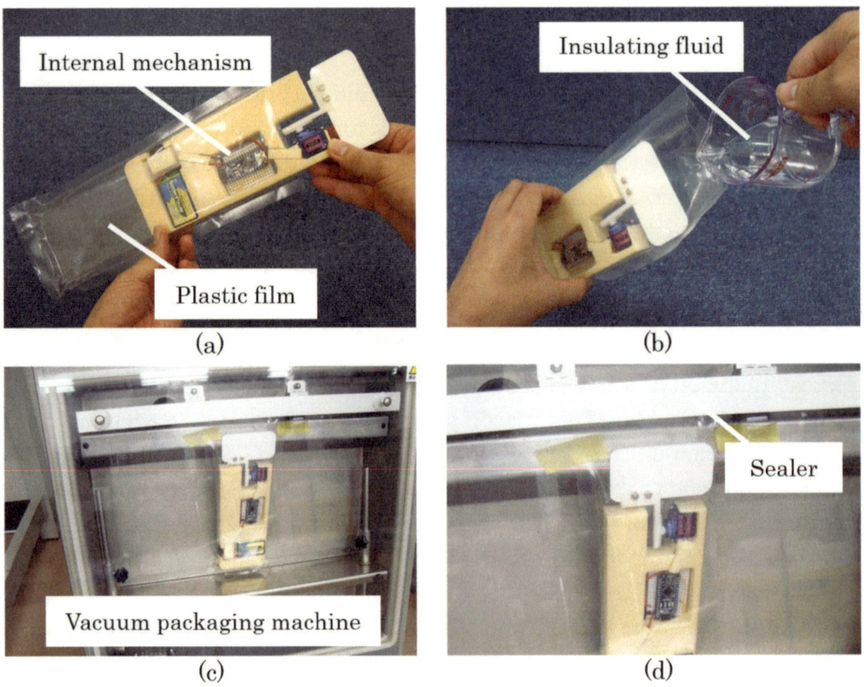

Figure 3. Fabrication process by robot packaging.

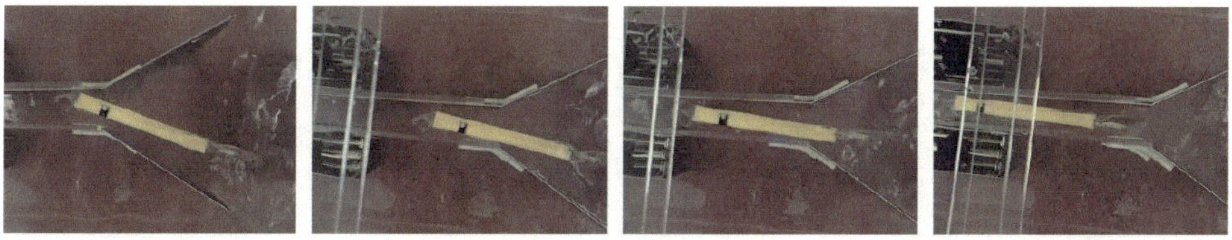

(a) (b) (c) (d)

Figure 4. Swimming test of our prototype at narrow space.

readily fabricate the entire body of a serial link robot laminated with a plastic film at low cost and in a short time. The body can also be lighter than other underwater systems.

2.2. Fish-like robot laminated with a plastic film

As shown in **Figure 2**, we proposed a fish-like robot as a prototype hardware laminated using a plastic film, which was fabricated using a vacuum packaging machine. In the prototype robot, we applied an insulating fluid (Fluorinert FC-3283; 3 M Corp.) filled in the body. The insulating fluid is generally used as a cleaning fluid for semiconductors. The specific gravity of the insulating fluid is approximately 1.83, which is heavier than water. Therefore, we used a copolymer foam (NiGK Corp.) not only for the frame structure of the body but also for generating buoyancy. The specific gravity of the foam is approximately 0.2, which is a much lighter fluid than water.

We also applied a servomotor (SG51R; Tower Pro) to the prototype robot actuator. In this prototype, we used a microcontroller (Arduino Nano ver. 3.1). A styrene board of 80 mm height, 40 mm width, and 2 mm thickness was used as an oscillation plate to generate the thrust force of the prototype robot in water. The plastic film (poly bag TL12-38; Fukusuke Kogyo Co. Ltd.) covering the internal contents of the robot was multilayered for use in food packaging. This film is sealed by thermal adhesion of a vacuum packaging machine. We made use of a TM-HV made by Furukawa Mfg. Co., Ltd. as a vacuum packaging machine. In this prototype, the depressurization time is 15 s. The sealing time is 4.0 s. A battery (9 V) is used as the power source for driving the electrical circuits and servomotor.

After design, the prototype size without the film was approximately 185 mm long, 80 mm high, and 18 mm wide. After fabrication, the prototype including the film in **Figure 2** was 230 mm long, 90 mm high, and 18 mm wide. Additionally, it weighed approximately 290 g, including 80 ml of the insulating fluid, thereby realizing almost neutral buoyancy of the body. Under these conditions, the oscillation plate driven by the servo motor was moved 10 deg. in the left and right directions at a frequency of approximately 2.0 Hz (**Figure 4**).

3. Modeling

This section presents a discussion of the derivation of the equations of motion of a serial link robot laminated with a plastic film. The laminated robot consists of a rigid link fin, a plastic film, and an enclosed insulating fluid as illustrated in **Figure 5**. In this section, we assume that

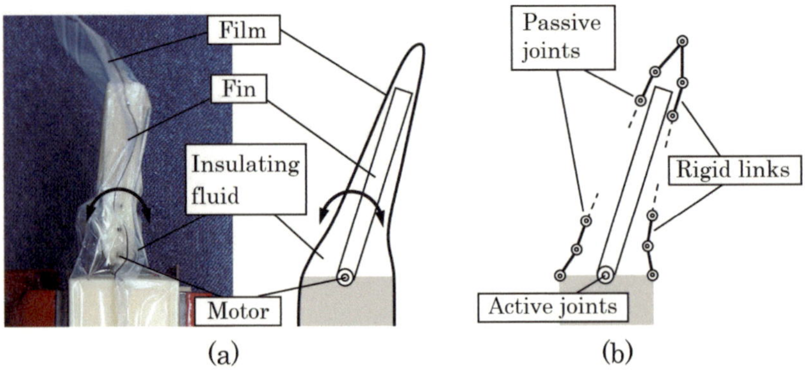

Figure 5. Modeling of a serial link robot laminated with a plastic film: (a) serial link robot laminated with a plastic film; (b) plastic film is modeled mathematically as a closed-loop link structure with multi-DOFs consisting of passive joints.

the plastic film is modeled as a multi-DOF closed-loop structure with passive joints that are actuated indirectly by a motor mounted on the base of the fin.

3.1. Rigid fin

The equation of fin motion in **Figure 5** is expressed simply as shown below.

$$I\ddot{\theta} = \tau \tag{1}$$

In this equation, scalar I represents the inertia of the rigid fin and the enclosed fluid, θ stands for the active joint angle of the fin, and the scalar τ includes an actuator torque and external torque related to the contact force from the film and additional inertia of the enclosed fluid.

To simplify the mathematical model of the contact force between the fin and the plastic film, we introduce a penalty method [42]. The penalty method treats elasticity and damping force for slight penetration between two objects. Here, we assume a small penetration between the fin and the joints of the links, as shown in **Figure 6(a)**. Contact force f_{pi} based on the penalty method is calculated as

$$f_{pi} = K_{pi}D_i - B_{pi}\dot{D}_i \tag{2}$$

where D_i denotes the small penetration between the fin and the plastic film at a contact point. K_{pi} and B_{pi} represent the elasticity and damping constant coefficients, respectively. As a result, the external torque τ_p generated by the contact forces f_{pi} can be calculated as

$$\tau_p = \sum_i \left(l_i \times f_{pi}\right) \tag{3}$$

where vector l_i represents the position for contact force f_{pi}.

In addition, the inertia of the enclosed fluid is included as well as the rigid inertia in the mathematical model because the rigid link fin carries the enclosed fluid during the fin motion.

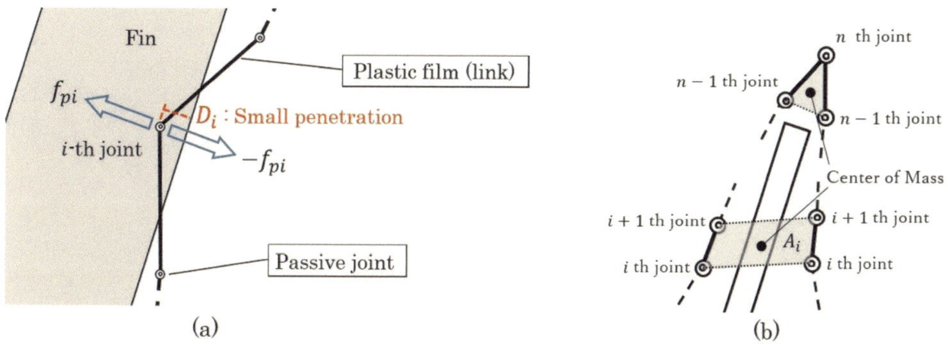

Figure 6. (a) Contact force between the fin and the plastic film based on the penalty method, and (b) calculation of the additional inertia of the enclosed fluid in the plastic film.

As illustrated in **Figure 6(b)**, a small area A_i of a trapezoid is determined by four points (or a triangle determined by three points at the tip of the fin,) and the center of the mass is used to estimate the fluid inertia. Mass M_{Fi} and inertia I_{Fi} of the fluid in a small area are obtained quantitatively as

$$M_{Fi} = A_i d\rho, I_{Fi} = M_{Fi} l_{Fi}^2 \tag{4}$$

where d, ρ, and l_{Fi} represent the depth of area A_i, the density of the insulating fluid, and the moment arm for the mass M_{Fi}, respectively. Consequently, the equations of the fin motion can be rewritten as follows.

$$(I_R + I_F)\ddot{\theta} = \tau_A + \tau_p \tag{5}$$

Therein, I_R signifies the inertia of the rigid fin. $I_F = \sum_j I_{Fi}$, τ_A stands for the actuator torque.

3.2. Plastic film

In this chapter, the plastic file is modeled as multi-DOF links with passive joints and with a closed kinematic loop constraint as illustrated in **Figure 7**. We use a method based on Lagrange-D'Alembert formulation on reduced system described in an earlier report [43, 44] to derive equations of motion of the closed-loop links.

First, a link of one of the closed-loop links, that is, is cut and a tree system (two serial link structures) is formed, as portrayed in **Figure 7**. Equations of motion of the serial link structures can be obtained easily using Newton-Euler method or Direct Lagrangian method. Because a plastic film is generally very light, the mass of the links can be negligible. Therefore, the equations of the motion of the multi-DOF serial link structures are only represented by passive joints with elasticity and damping effects:

$$B_R\dot{\theta}_R + K_R\Delta\theta_R = \tau_R \tag{6}$$

$$B_L\dot{\theta}_L + K_L\Delta\theta_L = \tau_L \tag{7}$$

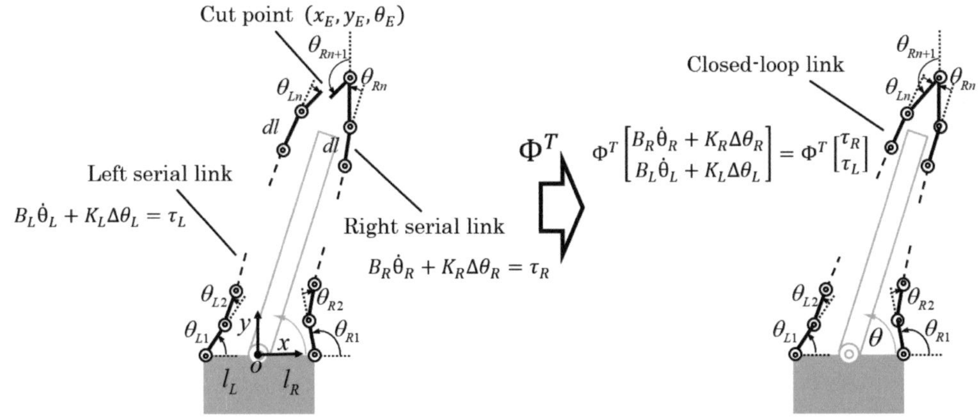

Figure 7. Mathematical model of the closed-loop link structure for a plastic film.

Therein, vectors $\theta_R = [\theta_{R1}, \cdots, \theta_{Rn+1}]^T$ and $\theta_L = [\theta_{L1}, \cdots, \theta_{Ln}]^T$ denote the passive joint angle vectors of the film, the vectors $\Delta\theta_R = \left[\tan\frac{\Delta\theta_{R1}}{2}, \cdots, \tan\frac{\Delta\theta_{Rn+1}}{2}\right]^T$ and $\Delta\theta_L = \left[\tan\frac{\Delta\theta_{L1}}{2}, \cdots, \tan\frac{\Delta\theta_{Ln}}{2}\right]^T$, where $\Delta\theta_{Ri}$ and $\Delta\theta_{Li}$ represent the rotational displacement for a joint. The matrices $B_R = \text{diag}(b_{R1}, \cdots, b_{Rn+1})$ and $B_L = \text{diag}(b_{L1}, \cdots, b_{Ln})$ represent the damping matrices, $K_R = \text{diag}(k_{R1}, \cdots, k_{Rn+1})$ and $K_L = \text{diag}(k_{L1}, \cdots, k_{Ln})$ are constant matrices related to the flexural rigidity and curvature of the plastic film. τ_R and τ_L represent the external torque related to the hydrodynamic effect acting on the film and the contact force between the rigid fin and the film, respectively.

Based on reports of the literature [43, 44], the equations of motion of the closed-loop structure is obtainable as

$$\Phi^T \begin{bmatrix} B_R\dot{\theta}_R + K_R\Delta\theta_R \\ B_L\dot{\theta}_L + K_L\Delta\theta_L \end{bmatrix} = \Phi^T \begin{bmatrix} \tau_R \\ \tau_L \end{bmatrix} \tag{8}$$

where Φ^T is the $(2n-2) \times (2n+1)$ matrix which comprises the Jacobian matrix of the constraint equations. Matrix Φ^T reduces the number of the equations from $(2n+1)$ to $(2n-2)$.

The following part of this discussion explains the external torque τ_R and τ_L. When forces f_i apply to their links, the torque is given as the following.

$$\begin{bmatrix} \tau_R \\ \tau_L \end{bmatrix} = \begin{bmatrix} \sum_i J_{Ri}^T(\theta_{Ri}) f_{Ri} \\ \sum_j J_{Lj}^T(\theta_{Lj}) f_{Lj} \end{bmatrix} \tag{9}$$

In that equation, J_{Ri} and J_{Li} are the Jacobian matrices of a contact point where a force is applied. Here, we consider forces related to the hydrodynamic effects and the contact forces between the plastic film and the rigid fin.

As described before, the contact force between the rigid fin and the film is derived based on the penalty method. The force applied to the film is expressed as

$$f'_{pi} = -f_{pi} = -\left(K_p D - B_p \dot{D}\right) \tag{10}$$

where f'_{pi} is the reaction force of f_{pi} in **Figure 6(a)**. Using Eq. (9), the torque related to the interference between the film and the fin is

$$\begin{bmatrix} \tau_{pR} \\ \tau_{pL} \end{bmatrix} = \begin{bmatrix} \sum_i J_{Ri}^T(\theta_{Ri}) f'_{piR} \\ \sum_j J_{Lj}^T(\theta_{Lj}) f'_{pjL} \end{bmatrix} \tag{11}$$

Finally, we assume that the hydrodynamic damping force acts as an external force on each link of the film. The hydrodynamic force acting on a link is modeled as shown below (**Figure 8**):

$$f_{Di} = \frac{1}{2} C_d \rho_w S_i |v_{\perp i}| v_{\perp i} \tag{12}$$

In this equation, C_d represents the drag force coefficient, ρ_w stands for the surrounding water density outside the film, S_i is the representative area of the ith link, and $v_{\perp i}$ is the relative flow velocity that is perpendicular to the ith link. We presume that the film moves in still water. The velocity of a link can be approximated by the following.

$$v_{\perp i} = R_i(\theta_i) J_i^T(\theta_i) \dot{\theta}_i \tag{13}$$

For vector $\theta_i = [\theta_{R1}, \cdots, \theta_{Ri}]^T$ or $\theta_i = [\theta_{L1}, \cdots, \theta_{Li}]^T$, matrix R_i is the rotational matrix to calculate the velocity perpendicular to the link. J_i is the Jacobian matrix of the ith link. However, we assume that $v_{\perp i} = 0$ $\left(f_{Di} = 0\right)$ when the ith link of the plastic film moves toward the rigid fin in the insulating fluid because the encapsulated insulating fluid in the plastic film moves with the fin and the film. Therefore, the relative velocity $v_{\perp i}$ between the film and the insulating fluid is almost zero.

Consequently, the external torque related to the contact forces and the hydrodynamic forces is summarized as shown below.

$$\begin{bmatrix} \tau_R \\ \tau_L \end{bmatrix} = \begin{bmatrix} \sum_i J_{Ri}^T(\theta_{Ri}) f'_{piR} + \sum_k J_{Rk}^T(\theta_{Rk}) f_{DkR} \\ \sum_j J_{Lj}^T(\theta_{Lj}) f'_{pjL} + \sum_k J_{Rk}^T(\theta_{Rk}) f_{DkR} \end{bmatrix} \tag{14}$$

In the model shown in **Figure 5**, three constraint equations related to the position $\left[x_E, y_E\right]^T$ and orientation θ_E of the two serial link structures should be considered. The constraint reduces the degrees of freedom of the serial link structures from $2n + 1$ to $2n - 2$. In other words, when configuration singularity does not occur, angular velocities $\left[\dot{\theta}_i, \dot{\theta}_j, \dot{\theta}_k\right]^T$ of any three joints of the closed-loop structure can be expressed by angular velocities of the $2n - 2$ other joints.

$$\left[\dot{\theta}_i, \dot{\theta}_j, \dot{\theta}_k\right]^T = \mathrm{J}(\theta_R, \theta_L)\dot{q} \tag{15}$$

Therein, $\dot{q}$ is the $2n - 2$ vector composed of all joint angular velocities except for $\dot{\theta}_i$, $\dot{\theta}_j$, $\dot{\theta}_k$ and $\mathrm{J}(\theta_R, \theta_L)$ is the $3 \times (2n - 2)$ matrix derived from the constraint equations.

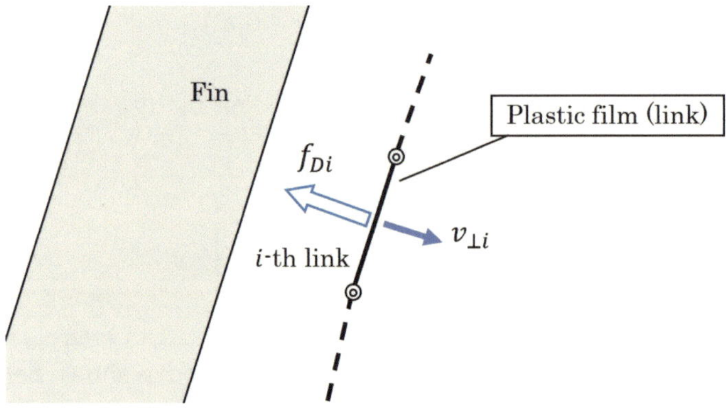

Figure 8. Model of the hydrodynamic damping force acting on the ith link.

Consequently, the equations of film motion are rewritten as shown below.

$$\left[\mathrm{B}' + J^T(\theta_R, \theta_L)\mathrm{B}''\mathrm{J}(\theta_R, \theta_L)\right]\dot{q} + \Phi^T\begin{bmatrix} K_R\Delta\theta_R \\ K_L\Delta\theta_L \end{bmatrix} = \Phi^T\begin{bmatrix} \tau_R \\ \tau_L \end{bmatrix} \tag{16}$$

Therein,

$$\mathrm{B}' = \begin{bmatrix} B'_R & 0 \\ 0 & B'_L \end{bmatrix}, \mathrm{B}'' = \mathrm{diag}\{b_i, b_j, b_k\} \tag{17}$$

where B'_R and B'_L denote the $(2n-2) \times (2n-2)$ matrices excluded the i, j, kth rows and columns from B_R and B_L, respectively . The characters b_i, b_j, b_k are the damping coefficients for the i, j, kth joints.

We use Eqs. (5) and (16) of the rigid fin and film motion for numerical simulation of the motion of the fin robot laminated by the plastic film.

4. Simulation

This section presents numerical simulation of the two-dimensional motion of the fin robot laminated by the plastic film. The purpose of the simulator is to design the laminated robots and to estimate their motion performance. A graphical simulator was developed for this purpose using software (Visual Studio 2010; Microsoft Corp.) and OpenGL library. The code was written in C language. The derived equations of the motion were solved numerically using the Runge-Kutta-Gill method by which the time step size was 0.001 s.

4.1. Numerical conditions

Table 1 presents the physical parameters of the laminated robot. We investigated the fin motion in three flexural rigidities of the plastic film. One was the value of the flexural rigidity of the film (3.14 gf cm^2) used to fabricate the fish robot (**Figure 2**). The other two values were 10

Moment of inertia of the fin [kg m^2]	I_R	0.0004033
b_{Ri} and b_{Li} in Eqs. (6) and (7)	b_{Ri}, b_{Li}	0.004
k_{Ri} and k_{Li} in Eqs. (6) and (7)	k_{Ri}, k_{Li}	0.001231
Position of the right 1st joint	$(l_R, 0)$	$(0.001, 0)$
Position of the left 1st joint	$(l_L, 0)$	$(-0.001, 0)$
Link length of the fin [m]	dl	0.005
Fin depth [m]	d	0.1
Density of insulating fluid [kg/m^3]	ρ	1830
Water density [kg/m^3]	ρ_W	1000
Representative area [m^2]	$S_i = d \times dl$	0.0005
Drag coefficient	C_D	1.3

Table 1. Parameters used for simulation.

times and 100 times the value of 3.14 (31.4 gf cm^2 and 314 gf cm^2). The plastic film had 41 passive joints to imitate a flexible film. The initial angles of the entire joints were set at 0 rad except for the following.

$$\theta_{R1} = \frac{92.86241\pi}{180}, \theta_{R17} = \frac{14.59520\pi}{180}, \theta_{R21} = \frac{145.08480\pi}{180}, \theta_{L1} = \frac{87.13760\pi}{180}, \theta_{L17} = -\frac{14.59520\pi}{180} \quad (18)$$

Based on the penalty method, the contact forces f_{pi} between the rigid fin and the entire joints of the plastic film were calculated using $f_{pi} = K_{pi}D_i - B_{pi}\dot{D}_i$ where $K_{pi} = 300$ and $B_{pi} = 0.1$ in the simulation.

The desired angle $\theta_d(t)$ [rad] for the rigid fin motion was given as

$$\theta_d(t) = \frac{\pi}{2} - \frac{\pi}{18}\sin(2\pi f t) \quad (19)$$

Frequency f of the fin motion was set at 2 Hz. For this desired angle, we made use of a conventional Proportional Derivative (PD) feedback controller $\tau_A = K_P[\theta_d(t) - \theta(t)] - K_D\dot{\theta}$ where $K_P = 30$ and $K_D = 0.1$. We assumed that a commercial waterproof servomotor (SG51R; Tower Pro) was used for the fin motion and assumed that the actuator had small lower and upper torque limits: $-0.0588 < \tau < 0.0588$ [Nm].

We presume that the relative flow velocity $v_{\perp i}$ for the calculation of the hydrodynamic force can be expressed approximately by the angular velocities $\dot{\theta}_R$ and $\dot{\theta}_L$ and joint angles θ_R and θ_L. There is no disturbance of water flow in the simulation.

4.2. Numerical results

We conducted a numerical simulation to investigate the fin motion of different flexural rigidities of the plastic film. **Figure 9** presents an illustration of how the fin changes in time series in the cases of (a) 3.14, (b) 31.4, and (c) 314 gf cm^2. **Figures 10** and **11** portray plots of the tracking

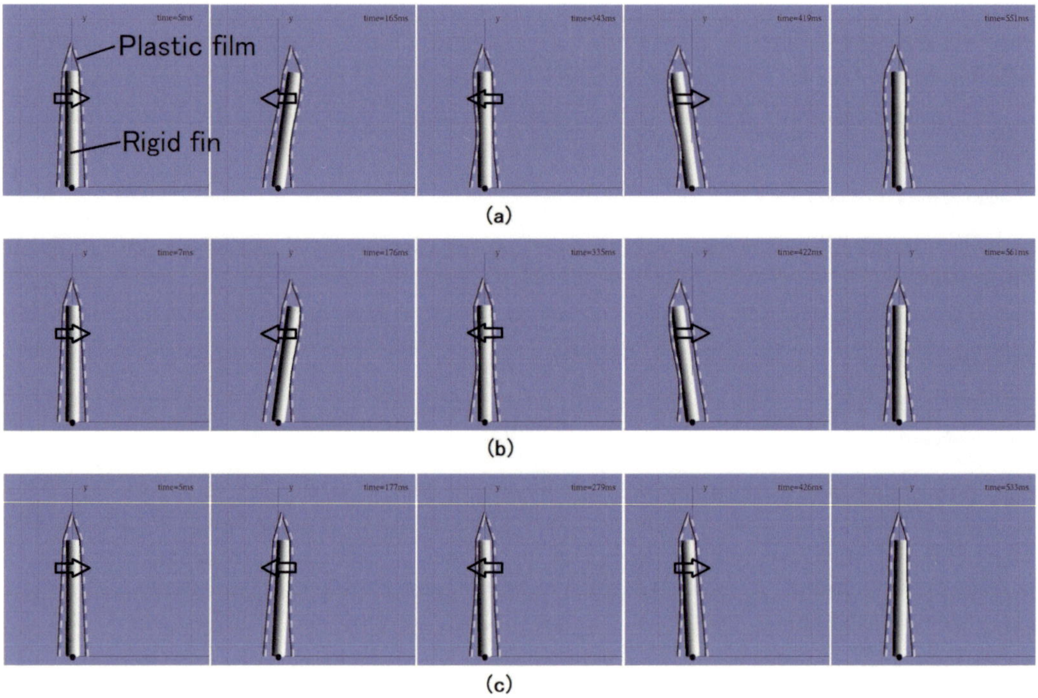

Figure 9. Two-dimensional simulation in three flexural rigidities of the plastic film. (a) 3.14, (b) 31.4, and (c) 314 gf cm^2.

data for the desired trajectory in Eq. (19) and the torque patterns in each case. For the value of 3.14 gf cm^2 in (a) that was the actual value for the prototype robot, the fin achieved smoothly reciprocating motion in the range of $\pm$ 8 degrees at 2 Hz in the simulation. The actual prototype robot in **Figure 4** also achieved approximately 10-deg. reciprocating motion at 2 Hz in the tank test. In (b), the flexural rigidity was 10 times different, no great difference in performance was found between the case (a) and the case (b). Result in (c) shows that higher flexural rigidity tended to prevent the fin motion. These numerical results demonstrate that the low torque actuator can inflect the robot body laminated by the thin plastic film with lower flexural rigidity.

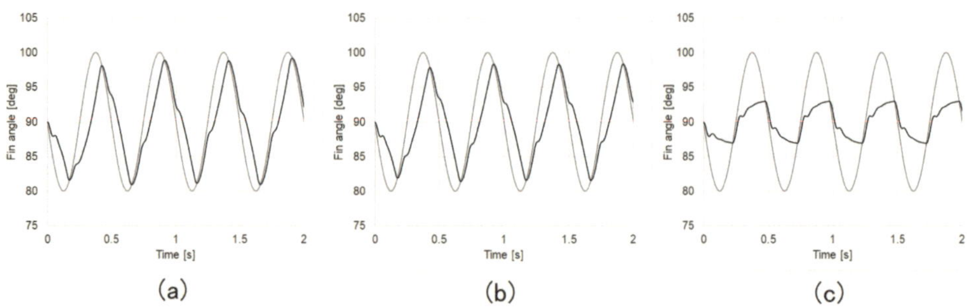

Figure 10. Tracking performance for the desired fin angle. (a) 3.14, (b) 31.4, and (c) 314 gf cm^2.

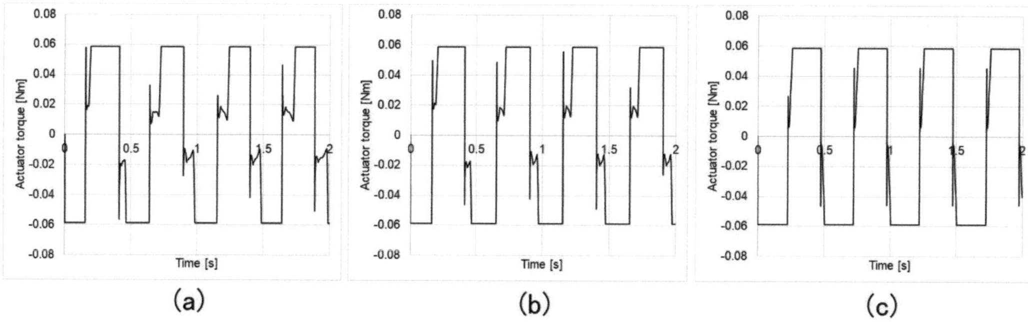

Figure 11. Torque pattern during the fin motion. (a) 3.14, (b) 31.4, and (c) 314 gf cm^2.

5. Conclusion

This chapter described the equations of motion of serial link robots laminated with a plastic film. We have proposed robot packaging method to improve waterproofing and dustproofing of serial link robots. To improve lubrication between the links and the film, we encapsulated an insulating fluid in the plastic film. Considering these conditions, we derived the equations of motion of the laminated robot to be useful for hardware design, motion analysis, and performance improvement such as thrust force and energy efficiency. In the derivation of the equations of motion, we assumed the plastic film as a closed-loop link structure with passive joints. Through numerical simulation based on the derived mathematical model of the fish-like robot, we estimated the motion performance of the fin in different flexural rigidities. We confirmed that the low torque actuator can inflect the laminated body because of the thin plastic film with lower flexural rigidity. Future work includes design of robots laminated with a plastic film using our mathematical model.

Acknowledgements

This work was partially supported by Furukawa Mfg. Co. Ltd. and partially by the Center of Innovation Program from Japan Science and Technology (JST) Agency. This work was also partially supported by JSPS KAKENHI (Grant Number 15 K18011).

Author details

Norimitsu Sakagami[1]* and Mizuho Shibata[2]

*Address all correspondence to: sakagami@scc.u-tokai.ac.jp

1 Department of Navigation and Ocean Engineering, School of Marine Science and Technology, Tokai University, Japan

2 Department of Robotics, Faculty of Engineering, Kindai University, Japan

References

[1] Lozano-Perez T, Jones JL, Mazer E, O'Donnell PA. Task-level planning of pick-and-place robot motions. Computer. 1989;**22**(3):21-29

[2] Wang D, Vidyasagar M. Modeling a class of multilink manipulators with the last link flexible. Transactions on Robotics and Automation. 1992;**8**(1):33-41

[3] Sicard P, Levine MD. An approach to an expert robot welding system. Transactions on Systems, Man, and Cybernetics. 1988;**18**(2):204-222

[4] Liu Y, Zhang Y. Toward welding robot with human knowledge: A remotely-controlled approach. Transactions on automation science and Engineering. 2015;**12**(2):769-774

[5] Siciliano B, Khatib O. Handbook of Robotics. 2nd ed. Spring; 2017

[6] Neo ES, Yokoi K, Kajita S, Kanehiro F, Tanie KA. Switching command-based whole-body operation method for humanoid robots. Transactions on Mechatronics. 2005;**10**(5):546-559

[7] Guan Y, Neo ES, Yokoi K, Tanie K. Stepping over obstacles with humanoid robots. Transactions on Robotics. 2006;**22**(5):958-973

[8] Bicchi A. Hands for dexterous manipulation and robust grasping: A difficult road toward simplicity. Transactions on Robotics and Automation. 2000;**16**(6):652-662

[9] Kawasaki H, Komatsu T, Uchiyama K. Dexterous anthropomorphic robot hand with distributed tactile sensor: Gifu hand II. Transactions on Mechatronics. 2002;**7**(3):296-303

[10] Huang Q, Yokoi K, Kajita S, Kaneko K, Arai H, Koyachi N, Tanie K. Planning walking patterns for a biped robot. Transactions on Robotics and Automation. 2001;**17**(3):280-289

[11] Grizzle JW, Abba G, Plestan F. Asymptotically stable walking for biped robots: Analysis via systems with impulse effects. Transactions on Automatic Control. 2001;**46**(1):51-64

[12] Ma S, Tomiyama T, Wada H. Omnidirectional static walking of a quadruped robot. Transactions on Robotics. 2005;**21**(2):152-161

[13] Hirose S, Fukuda Y, Yoneda K, Nagakubo A, Tsukagoshi H, Arikawa K, Endo G, Doi T, Hodoshima R. Quadruped walking robots at Tokyo Institute of Technology. Robotics & Automation Magazine. 2009;**16**(2):104-114

[14] Hirose S, Yamada H. Snake-like robots [Tutorial]. Robotics & Automation Magazine. 2009;**16**(1):88-98

[15] Tanaka M, Tanaka K. Shape control of a snake robot with joint limit and self-collision avoidance. Transactions on Control Systems Technology. 2017;**25**(4):1441-1448

[16] Conte J, Modarres-Sadeghi Y, Watts M, Hover FS, Triantafyllou MSA. Fast-starting mechanical fish that accelerates at 40 m/s2. Bioinspiration and Biomimetics. 2010;**5**(3):035004

[17] Yu J, Tan M, Wang L. Cooperative control of multiple biomimetic robotic fish. Recent Advances in Multi Robot Systems. 2008;**14**:263-290

[18] Higashimori M, Harada M, Ishii I, Kaneko M. Torque pattern generation towards the maximum jump height. In: Proceedings of IEEE International Conference on Robotics and Automation 2006. 1096-1101

[19] Craig JJ. Introduction to Robotics: Mechanics and Control. 3rd ed. Pearson.; 2004

[20] Eriksen CC, Osse TJ, Light RD, Wen T, Lehman TW, Sabin PL, Ballard JW, Chiodi AM. Seaglider: A long-range autonomous underwater vehicle for oceanographic research. Journal of Oceanic Engineering. 2001;**26**(4):424-436

[21] Sherman J, Davis RE, Owens WB, Valdes J. The autonomous underwater glider "spray". Journal of Oceanic Engineering. 2001;**26**(4):437-446

[22] Webb DC, Simonetti PJ, Jones CP. SLOCUM: An underwater glider propelled by environmental energy. Journal of Oceanic Engineering. 2001;**26**(4):447-452

[23] White DA. Modular design of Li-ion and Li-polymer batteries for undersea environments. Marine Technology Society Journal. 2009;**43**(5):115-122

[24] Shibata M, Sakagami N. Fabrication of a fish-like underwater robot with flexible plastic film body. Advanced Robotics. 2015;**29**(1):103-113

[25] Shibata M. Fish-like robot encapsulated by a plastic film. Recent Advances in Robotic Systems. Wang G editor. InTech. 2016:235-251. DOI: 10.5772/63506. Available from: https://www.intechopen.com/books/recent-advances-in-robotic-systems/fish-like-robot-encapsulated-by-a-plastic-film

[26] Zhang Y, Cheng MC, Pillay P. A novel hysteresis core loss model for magnetic laminations. Transactions on Energy Conversion. 2011;**26**(4):993-999

[27] Baghel APS, Chwastek K, Kulkarni SV. Modelling of minor hysteresis loops in rolling and transverse directions of grain-oriented laminations. Electric Power Applications. 2015;**9**(4):344-348

[28] Rasilo P, Singh D, Aydin U, Martin F, Kouhia R, Belahcen A, Arkkio A. Modeling of hysteresis losses in ferromagnetic laminations under mechanical stress. Transactions on Magnetics. 2016;**52**(3):7300204

[29] Khang HV, Arkkio A. Eddy-current loss modeling for a form-wound induction motor using circuit model. Transactions on Magnetics. 2012;**48**(2):1059-1062

[30] Hamzehbahmani H, Anderson P, Hall J, Fox D. Eddy current loss estimation of edge Burr-affected magnetic laminations based on equivalent electrical network part I: Fundamental concepts and FEM Modeling. Transactions on Power Delivery. 2014;**29**(2):642-650

[31] Chen J, Wang D, Cheng S, Wang Y, Zhu Y, Liu Q. Modeling of temperature effects on magnetic property of nonoriented silicon steel lamination. Transactions on Magnetics. 2015;**51**(11):2002804

[32] Petrun M, Steentjes S, Hameyer K, Dolinar D. 1-D lamination models for calculating the magnetization dynamics in non-oriented soft magnetic steel sheets. Transactions on Magnetics. 2016;**52**(3):7002904

[33] Jensen BB, Guest ED, Mecrow BC. Modeling overlapping laminations in magnetic core materials using 2-D finite-element analysis. Transactions on Magnetics. 2015;**51**(6):7403006

[34] Zheng W, Cheng Z. An inner-constrained separation technique for 3-D finite-element modeling of grain-oriented silicon steel laminations. Transactions on Magnetics. 2012;**48**(8): 2277-2283

[35] Sumoto H, Yamaguchi S. Development of a motion control system using phototaxis for a fish type robot In: Proceedings of the International Conference on Offshore and Polar Engineering 2010. pp. 307-310

[36] Kobayashi K, Yoshikai T, Inaba M. Development of humanoid with distributed soft flesh and shock-resistive joint mechanism for self-protective behaviors in impact from falling down In: Proceedings of International Conference on Robotics and Biomimetics 2011. pp. 2390-2396

[37] Hayashi M, Sagisaka T, Ishizaka Y, Yoshikai T, Inaba M. Development of functional whole-body flesh with distributed three-axis force sensors to enable close interaction by humanoids. In: Proceedings of International Conference on Intelligent Robots and Systems 2007. pp. 3610-3615

[38] Aono T, Nakamura Y. Design of humanoid with insert-molded cover towards the variety of exterior design of robots. In: Proceedings of International Conference on Intelligent Robots and Systems 2005. pp. 3342-3347

[39] Erturk A. Macro-fiber composite actuated piezoelectric robotic fish. In: Robot Fish: Bio-inspired Fishlike Underwater Robots 2015. pp. 255-283

[40] Shintake J, Shea H, Floreano D. Biomimetic underwater robots based on dielectric elastomer actuators. In: Proceedings of International Conference on Intelligent Robots and Systems 2016. 4957-4962

[41] Ahvenainen R. Novel Food Packaging Techniques. Woodhead Publishing.; 2003

[42] Tamamoto T, Koganezawa K. Multi-joint gripper with stiffness adjuster. In: Proceedings of 2013 IEEE/RSJ International Conference on Intelligent Robots and Systems (IROS). 2013. pp. 5481-5486

[43] Yiu YK, Cheng H, Xiong ZH, Liu GF, Li ZX. On the dynamics of parallel manipulators. In: Proceedings of the IEEE International Conference on Robotics and Automation. 2001. pp. 3766-3771

[44] Sakagami N, Otomasu K, Choi SK. Modeling and simulation of closed-loop mechanical systems for underwater applications. In: Proceedings of the 47th ISCIE International Symposium on Stochastic Systems Theory and Its Applications. 2015. pp. 247-252

Chapter 5

Advanced Technologies in Manufacturing 3D-Layered Structures for Defense and Aerospace

Dionysios E. Mouzakis

Additional information is available at the end of the chapter

http://dx.doi.org/10.5772/intechopen.74331

Abstract

In the past 20 years, a great progress has been made in additive manufacturing techniques, which has led to numerous applications in aeronautical and defense structures. Though not all advanced materials and alloys, can be automatically layered by a rapid prototyping system or machine, several interesting application have seen the light of publicity in many sectors. Efforts are underway to apply the automated layering technologies in as many materials as possible, mostly nowadays plastics, reinforced-polymers, and metals can be processed by such systems in order to produce three-dimensional parts. The work is underway internationally in order to promote more and more applications of additive manufacturing or automated layering and to lower the costs in such systems. This paper aims at presenting a review of the additive manufacturing history presenting the major steps that lead to the explosion of this technology, and with a special focus on advanced 3D structures in aerospace and defense applications. An insight is also given on the four dimensions of manufacturing concept.

Keywords: laminated manufacturing, 3D printing, rapid prototyping, additive manufacturing

1. Early research and development of making 3D objects

It was back in the late 1960s, when the creation of solid objects using photopolymerized resins was attempted, by employing a laser. The attempt took place at the Battelle Memorial Institute in Columbus, Ohio, USA [1]. During this pioneering experiment, two laser beams with non-similar wavelengths were pointed to intersect in the middle of a transparent container filled with of resin. Inside the vat of resin (usual term nowadays), the polymer crosslinked and solidified at the point where the laser beams intersected. DuPont had already made available the photopolymerization of resins as a technology in the 1950s. Quite similar to this approach, Swainson filed a patent titled *"Method of Producing a 3D Figure by Holography"* in Denmark, in 1967. His procedure was also based on dual-laser beam scheme. From 1967 to 1994, the laser-induced photopolymerization was being developed, but until then no functional 3D printer was offered to the market. In the early 1970s, a company called Formigraphic Engine Co., applied the dual-laser intersection paten, aiming at the first commercially available laser-prototyping machine. At that time, they coded the methodology under the term "photochemical machining." That company, however, succeeded to present the creation of a 3D object in 1974. Later, Formigraphic became Omtec Replication, and led the development of similar techniques with the help of DARPA funding. As the decade expired, a number of patents regarding solid photography were claimed by Dynell Electronics Corp. Their invention described the cutting of cross sections, in any soft material available, usually carton or plastic, using either a milling machine or laser, guided by computer control and stacking the sections sequentially, to form a 3D object. Nowadays, the technique is called laminated object manufacturing (LOM). Dynell merged with United Technologies Corp. in 1977. Their merger was independent and was called Solid Photography. This company opened an affiliated retail outlet named Sculpture, which by mid-1981 had changed its name to Robotic Vision.

1.1. Stereolithography emerges

According to many sources, Hideo Kodama, working at the Nagoya Municipal Industrial Research Institute (Nagoya, Japan), was one of the first to invent the single-beam laser curing approach, In mid-1980, he filed a patent in Japan, which unfortunately expired without proceeding to the examination stage, this being a requirement of the Japanese patent application procedure. Kodama apparently had obstacles in securing funds for additional research and development. In 1981, he published a paper titled *"A Scheme for Three Dimensional Display by Automatic Fabrication of Three Dimensional Model"* [2] and another paper titled *"Automatic Method for Fabricating a Three-Dimensional Plastic Model With Photo-Hardening Polymer"* that outlined his work in detail and offering thus the technology openly to the public [3]. During his experiments, UV rays were projected, using a mercury lamp, (Toshiba) into a photosensitive-photopolymerizing resin called Tevistar® produced by Teijin. In Kodama's innovating technique, black and white film was used to mask and accurately determine the region of exposure to UV light, and thus to define each cross section. His first paper also described the application of an x-y plotter device and the use of optical fibers as carriers of UV light. His approach was used by CMET Company in the SOUP 530, 600, and 850 stereolithographic systems. In his paper, Kodama quotes, *"If the solidified layer is immersed into the liquid with the top at a depth equal to the thickness of the layer to be solidified, its top surface is covered with unsolidified liquid polymer,...."*, actually giving the description of the main core procedure of the stereolithography technology. Kodama's experiments, and papers, are nowadays considered to be the first successful application of working additive manufacturing (AM) techniques worldwide.

The real market-available additive manufacturing, however, first saw light in 1987 with stereolithography (SL) from 3D Systems. Their process involved solidification of thin layers of light-sensitive liquid resin polymer using a UV-laser beam. They marketed their SLA-1, being the first commercially available AM machine worldwide. This system was the precursor of the SLA 250 machine, which became a commercial success (SLA acronym means Stereo-Lithography-Apparatus). SLA 250 was replaced by Viper SLA, and nowadays replaced by the ProJet series of SLA Printers.

A year later, in 1988, 3D Systems and Ciba-Geigy formed a partnership aiming at the development of stereolithography materials and marketed their first-generation acrylate resins. DuPont developed its stereolithography system named Somos along with proper resins in the same year. Loctite had also attempted to enter the SL resins market in the late 1980s, but closed the corresponding department in 1993.

Japan's NTT Data CMET and Sony/D-MEC commercialized versions of stereolithography in 1988 and 1989, as an answer to 3D Systems SL in the U.S. NTT Data CMET (now a part of Teijin Seiki, a subsidiary of Nabtesco) called its machine Solid Object Ultraviolet Plotter (SOUP) and Sony/D-MEC (now D-MEC) called its device Solid Creation System (SCS). Sony closed its production for SL systems for D-MEC in 2007. In 1988, Asahi Denka Kogyo introduced the first epoxy resin for the CMET SL machine. In 1989, Japan Synthetic Rubber (now JSR Corp.) and DSM Desotech started to supply polymers for the Sony/D-MEC stereolithography machines.

In 1990, Electro Optical Systems (EOS) based in Germany sold its first Stereos stereolithography system. The same year, Quadrax introduced the Mark 1000 SL system, which employed visible light resin. The following year, Imperial Chemical Industries introduced a visible light resin product for the Mark 1000. ICI stopped selling its resin about 1 year later when Quadrax dissolved due legal issues with 3D Systems.

1.2. Layering systems-non SL

Back in 1991, three AM technologies were made available to the market, e.g., fused deposition modeling (FDM) as coined by Stratasys, solid ground curing (SGC) marketed by Cubital, and as mentioned above, laminated object manufacturing (LOM) by Helisys. FDM utilizes molten thermoplastic polymers in the form of a filament to structure an object in a layer-by-layer fashion. In SGC, a UV-sensitive liquid polymer is used, and each complete layer solidifies in each laser scan by ample UV light which passes through masks created using an electrostatic toner on a glass plate. LOM cut sheet material using a digitally guided laser and bonds the stacked layers together into a 3D object. Cubital and Helisys are not in the market anymore. Selective laser sintering (SLS) from DTM (now a part of 3D Systems) and the Soliform stereolithography system from Teijin Seiki became available in 1992. Using laser as an extreme heat source, SLS sinters powder materials by local fusion. DuPont had developed the Soliform technology originally, under the name Somos and licensed it to Teijin Seiki. The latter company had the exclusive distribution rights in parts of East Asia. Also, in the same year, a company called Allied Signal marketed vinylether Exactomer resin polymers for SL applications. In the next year, Soligen of Germany presented a device under the name direct shell production casting (DSPC). DSPC employed an inkjet printer mechanism, which deposited liquid binder onto ceramic powder. In this way, shells were formed for later use in the investment-casting procedure. The patent Soligen used was filed by the Massachusetts Institute of Technology (MIT). In January 2006, Solingen stopped its production of DSPC systems. That year, Denken marketed an SL system that featured a solid-state laser. It should be mentioned here that Denken's SL system was presumably, compact enough, to sit on a bench top. Moreover, it was also given away at a relatively low price, compared to other SL systems available in the market. In 1993, 3D Systems and Ciba made their first epoxy resin product for SLA, commercially available. In the same period, the QuickCast structuring scheme was presented. QuickCast, still used to-date, is a process in which hollow investment-casting molds are produced. After casting, the polymer mold would burn out, without damaging the fragile ceramic shell. In 1994, many new additive manufacturing systems found their way into the markets. ModelMaker from Solidscape (then called Sanders Prototype) was delivered, as many new systems from Japanese and European firms did. By making use of an inkjet print head, ModelMaker had the ability to deposit waxy materials layer, by layer. Aiming at a new market sector, namely jewelry makers, Meiko in Japan produced a novel small stereolithography device. Meiko is no more in the SL sector since 2006. Also, in Japan, KiraCorp. marketed its first non-stereolithographic device. The device called Solid Center was actually a complete LOM system featuring a typical laserprinter engine, toner, an x-y plotter, and knife; it was able to produce wood-like models by paper lamination. Kira referred to Solid Center as the first plain-paper 3D printer. In the same year 1994, Fockele & Schwarze (F&S) in Germany introduced a stereolithography machine, but on a limited basis, and a German company named EOS commercialized a printer called EOSINT based on laser-sintering technology. Ushio from Japan (Unirapid Inc. nowadays) sold its first stereolithography machine in 1995.

1.3. Low-cost 3D printers

In 1996, Stratasys introduced the Genisys machine. This type of printer utilized an extrusion process similar to FDM, however, based on technology developed at IBM's Watson Research Center. Being a decade almost in the market of stereolithography systems, 3D Systems offered to the market its first 3D printer (Actua 2100) in 1996, using a technology that deposits waxy materials in a layer after layer fashion, utilizing an inkjet-printing mechanism. In the same year, a company called Z Corp. marketed its Z402 modeling 3D printer, aiming at conceptual use. The Z402 machine profited from MIT's inkjet-printing (3DP) technology, and models were manufactured using starch- or plaster-based powders and a water-soluble liquid binder. Also in 1996, Schroff Development started to offer to the market its semi-automated paper lamination system below the threshold of $10,000. BPM Technology begun to sell its Personal Modeler 2100 model in 1996 too. By employing a process named ballistic particle manufacturing (BPM), the machine could deposit waxy material layers by use of an inkjet-printing head. The company ceased operations in October 1997. Kinergy based in Singapore started selling its Zippy paper lamination systems, which worked much alike as the LOM process. AeroMet founded in 1997 was a subsidiary of MTS Systems Corp. This company developed a procedure called laser additive manufacturing (LAM) that employed a high-power laser and titanium alloys in the form of powder. Until it stopped operations in December 2005, AeroMet was a 3D printed parts subcontractor for the aerospace industry. That year, Ciba acquired the Exactomer resins business from Allied Signal. In 1998, Yinhua Laser Rapid Prototypes Making & Mould Technology Co., Ltd. based in Beijing-China, intensively pursued the marketing of its products. Not by chance, Tsinghua University in Beijing, being the original developer of these systems, has developed processes much alike to FDM and other additive manufacturing technologies, since 1996. Autostrade started to market a stereolithography system called E-DARTS to firms in Japan at prices no higher than $25,000, in the same year. Also in 1998, Optomec made available to the industry its laser-engineered net shaping (LENS) metal powder system. The machine was based on the technology developed at Sandia National Labs. From that point, the markets begin to open even more and also the international demand for such systems. In March 1999, 3D Systems introduced a machine called ThermoJet, a much faster and less costly version of Actua 2100. At that time, 3D Systems was selling its SLA 7000 system for $800,000, which was the most expensive AM system for plastic materials, available worldwide. In April 1999, on demand by Motorola, a company under the name Extrude Hone AM business (nowadays named Ex One) installed its first ProMetal RTS-300 system, for building metal parts. The machine was utilizing MIT's 3DP inkjet-printing technology. In 1999, Fockele & Schwarze based in Germany, revealed its selective laser-melting system for steel-based powders. This system was developed in cooperation with the Fraunhofer Institute for Laser Technology, in Aachen. The same Institute provided know-how for Röders, who developed and sold its controlled metal buildup (CMB) machine. Also, in 1999, DSM acquired the Somos branch from DuPont. In January 2000, Helisys announced that Toyoda Machine Works of Japan would manufacture and sell LOM systems in Japan. In June 2000, Toyoda exhibited its proprietary machine based on LOM technology, at a technology fair in Tokyo. Sanders Design International proclaimed the development of a system named Rapid Tool Maker (RTM) in January 2000, and they also announced that it had licensed the RTM technology to a German company called Buss Modeling Technology (BMT). BMT, which was formerly Buss Müller Technology, had the strategic plan to manufacture RTM systems and provide it to the European markets. At that time, BMT announced manufacturing and marketing a color 3D printer based on powder and binder technology, developed by Aad van der Geest of the Netherlands. The process was quite similar to the 3DP process from Z Corp.

Since the beginning of the first decade of the new century, many systems are available in the market for producing layered structures, now under the name 3D printed structures utilizing the prevailing technologies as described above in summary and in the following chapters in detail. Nowadays, many affordable systems are available in the open market, even as DIY kits, for applications in almost all industrial sectors as well as for hobby and recreation as it will be shown.

2. Market placement

Up to year 2009, layering technology of 3D printing was largely employed for industrial applications, as reported above; however, exactly then, the patent protecting fused deposition modeling (FDM)—one of the most simple and commonly used 3D printing technologies nowadays—expired. Thanks to the RepRap people's project's mission and vision, to build a self-replicating 3D printer, the first desktop 3D printer was born. This caused an avalanche effect and many manufacturers followed, and the cost which was initially $200,000 back in early 2000, suddenly sunk below $2000, and the consumer 3D printing market took off in 2009. Nowadays, a simple DIY system costs about $150 for hobbyists.

The sales and market of 3D printing have been skyrocketing ever since. Of course, important patents on additive manufacturing expired and technology is made available to more bright minds, hence more innovations are foreseen in the near future as well as enormous revenues in the main and niece markets. There are, nowadays, almost 300,000 enterprise users and more than 1,000,000 private owners of 3D printers in the world. These numbers are doubling every other year. 3D printing industry is still in its childhood phase and its growing bigger and bigger. The number of companies that manufacture 3D printers has doubled in the last 2 years. According to Wohler's Report for 2017, 97 manufacturers produced and sold additive manufacturing machines and devices in 2016. A year later, they were 62. The industry achieved worldwide revenues of $6.063 billion in 2016 about three times the maximum forecasted value as seen in **Figure 1**.

2.1. 3D printing: advantages and disadvantages

It is of paramount significance to understand that 3D printing is a rapidly developing technology, rather immature, which comes with its family of inherent benefits, but also lags behind traditional manufacturing processes in many aspects. Examples from all aspects will enable the reader, get a grasp of these factors and foresee where the technology is headed in the near future.

3D printing allows designers and engineers create complex shapes and parts—many of which cannot be produced by traditional manufacturing techniques. Of course, evidently, manufacturing through additive methods implies that complexity comes at a price; elaborate product designs with complicated design features now cost just as much to produce as simple product designs that follow all the traditional rules of conventional manufacturing.

Utilizing traditional production methods, at a high volume or numbers of products, it is simply cheaper to make and sell products at reasonable prices to the consumer. Alternatively, 3D printing allows easy customization; one only needs to change the design digitally in CAD software to make changes with no additional tooling or other expensive or high efficiency

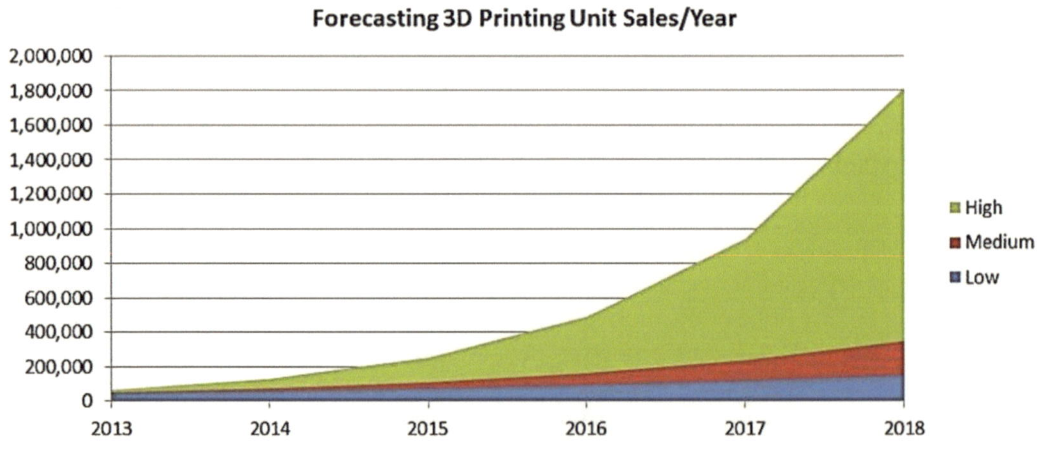

Figure 1. Estimated 3D printer sales per year from 2013 to 2018 [4].

manufacturing processes required to produce the final product. This results in an item that can be customized to meet a user's specific needs without additional manufacturing costs, only the design labor remaining.

In metal casting or injection molding, each specific part of each product requires a mold—a factor that can skyrocket manufacturing costs rapidly. To counteract these permanent manufacturing costs, most companies anticipate thousands of the same items being sold. On the other hand, 3D printing is *per se*, a "single tool" or maybe a zero-tool process. In this way, no tool is required and no need to change any condition of the process exists. Additionally, no hidden costs or lead times are involved in making an object complex or simple. Consequently, this paves the way for substantial decreases in production costs. At the cost of manufacturing time of course, needless to mention, as situation stands nowadays, because 3D printing is by no means considered a mass production technology.

Strategically thinking, once its manifested that there is no expensive tooling required to produce objects through 3D printing technologies, designers or entrepreneurs, might consider it, and they usually do so, as a cost effective method to produce items for a market test run or small production series. Possibly best, exploit the internet, through crowdfunding sites like Kickstarter, in order to launch their products. At the early stages of product development, it appears both also practical and wise to make design changes, in the product, without compromising their name in more formal—and expensive—manufacturing orders. Concluding, 3D printing technology opens a much less dire route to the market for those who want to materialize a novel product or an idea.

Most conventional manufacturing processes are subtractive especially the second grade ones: you start with a block of material (or a cast item), and usually through cutting, milling, drilling or similar, it is being processed at the intended final design. For many products—such as a bracket for an airplane—90% of the raw material is lost during processing.

On the other hand, 3D printing belongs to the additive processes; object is created from the raw material layer-by-layer. According to the laws of Mother Nature (she creates things exactly this way we found), when an object is manufactured this way, it only uses as much material—and energy—that is needed to create that particular object and no more. Additionally, most of these materials can be recycled and repurposed into more 3D-printed objects.

Alas, having all of the benefits of manufacturing through additive techniques, 3D printing is not yet competitive with conventional manufacturing processes, when it comes to large production volumes. The critical turning point lies between 1000 and 10,000 units, the numbers being a function on the material and the design. Of course, as the prices of printers and raw materials continue to decrease, the range of efficient production is expected to increase above the reported numbers in the following years.

Nowadays, there are more than 600 3D printing raw materials marketed, most of which are plastics and metals, but the choices are still limited compared to conventional product materials, available colors and finishes. The pseudo-lack of materials, however, is increasing, the number of new materials added to the 3D printing palette is growing rapidly now including wood, metals, alloys, composites, ceramics, and even chocolates.

A key aspect, of course, is the mechanical properties. In most 3D printing technologies, the part strength is neither uniform nor high, due to the layer-by-layer fabrication process. Practically, parts that have been 3D printed are usually weaker than their traditionally manufactured counterparts. Repeatability is also an open question; parts made on different 3D printers might have varying properties. However, as technical improvements are rapidly achieved and as novel continuous 3D printing processes like Carbon3D are made available, these disadvantages are prone to extinct in coming years.

Despite, the fact that we are not still able to manufacture 3D-printed objects with submicron tolerances like an iPhone, 3D printing technology is considered as a very straightforward and practical procedure of layering objects. These parts feature precision within the scale 20–100 microns, which correspond to a natural scale from the diameter of a human hair to the height of a single sheet of paper. 3D printing enables designers and engineers, who are creating objects with few tolerances and design details, to make products and bright ideas real. As known, many high-tech objects demand fine working parts and even finer details—such as the silent switch on the iPhone—it is still difficult to compete with the high precision capabilities of certain manufacturing processes, but time will prove this technology in every case. In every case, 3D printing is changing business model innovation in a very rapid manner (**Figure 2**) [7].

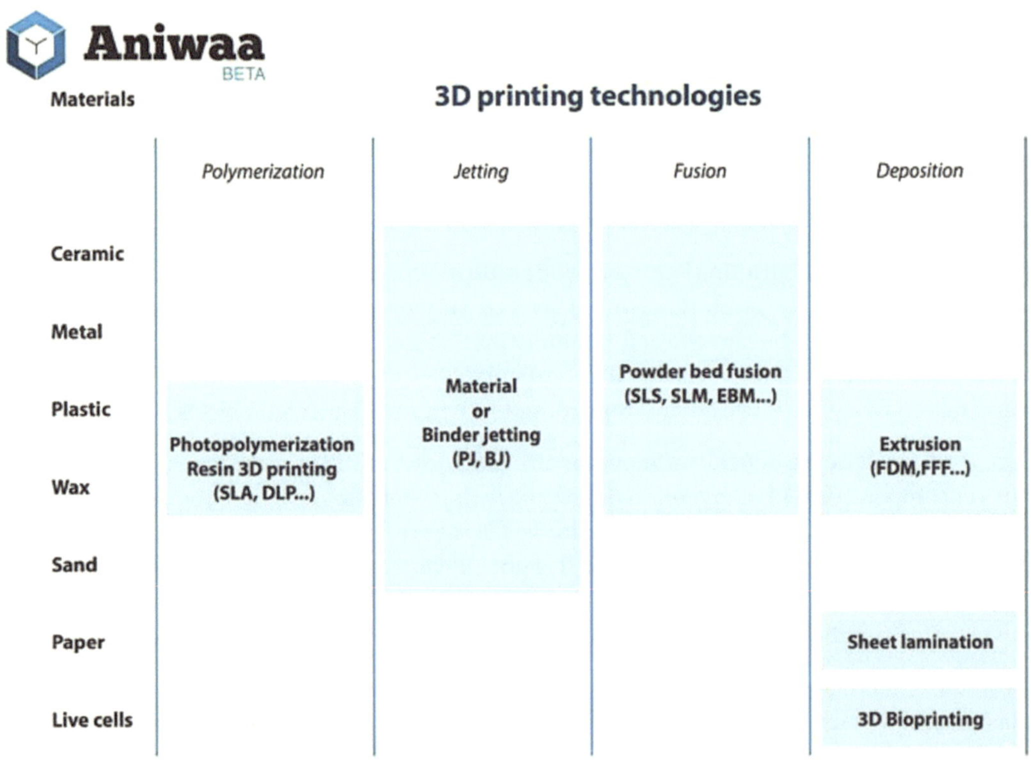

Figure 2. Available technologies and materials for 3D printing and lamination techniques. Source: www.aniwaa.com.

http://dx.doi.org/10.5772/intechopen.74331

3. 3D technologies review

This part refers to the main 3D printing technologies which enable us in printing layered structures in some detail. Only the principal ones in use nowadays are reported, and some innovative ones are shown in the applications chapter.

3.1. Fused deposition modeling (FDM)

The FDM printing process begins with a string of solid material called the filament. This line of filament is pulled from a reel attached to the 3D printer to a heated nozzle inside of the 3D printer that heats the material above its melting point. Once in a melted state, the material pushed out of a nozzle is extruded on a specific and predetermined path guided by the software on the computer usually instructed in G-code language. As the material is extruded, as a layer of the object on this path, it instantly cools down and solidifies—providing the base for the next layer of material until the entire object is manufactured.

Considered nowadays as the cheapest 3D printing technology commercially available, FDM also offers a wide variety of plastic-matrix materials in a rainbow of colors including ABS, PLA, nylon, and blends with more exotic materials, including carbon, bronze, or wood (**Figure 3**).

FDM is a considered to be the most practical choice for quick and low-cost prototyping. It can be used for a wide range of applications and objects with a typically wide palette of polymers as filaments in pure or reinforced form. Recently, FDM 3D printing has become very famous among hobbyists for enabling them to design and produce functional products, with embedded electronics and mechanical parts such as drones. FDM 3D printing is hampered by

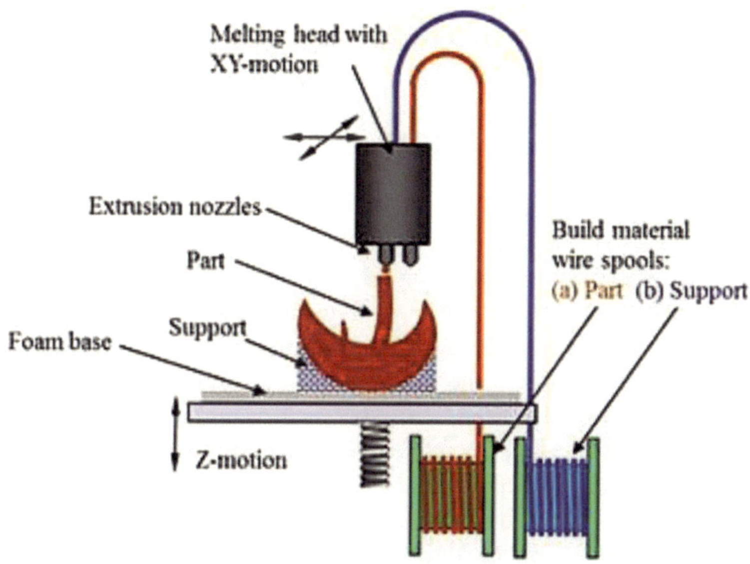

Figure 3. FDM technique sketch. Source: Caliskan and Durgun [5, 6].

design and material limitations, although improvements appear almost continuously nowadays. The technology generally is not considered suitable for more intricate designs or where high strength is required [8]. Usually, the parts manufactured with this technique can exhibit some internal anisotropy due to layering procedure [9].

3.2. Stereolithography and digital light processing (SLA & DLP)

These techniques are reported together, due to the fact that both technologies produce 3D-printed parts using a photo polymerizing polymer resin, featuring a UV light source to cross-link the liquid material [10].

Analytically, the procedure is as follows: a building platform submerges into a translucent tank (vat) filled with the liquid photo polymerizing resin. After submerging the tank, the UV light source located inside the machine, focuses through the bottom of the tank, scans each layer of the object, effectively solidifies-crosslinks, or polymerizes the material in other words. Consequently, the platform is lifted upward by a few microns, thus allowing a fresh layer of resin to flow beneath the object. The UV light source shall map and solidify the new layer onto the previous one. Micron-by-micron step, the process is repeated in a layer-by-layer fashion, top to bottom until the whole part is finished. The methods are differentiating only by the light source used: In SLA, a UV-laser is used; whereas in DLP, a UV-projector lamp is employed.

The progress made in the past decades delivered enabled 3D printing processes to be applied in desktop 3D printers. Needless to mention here, materials selection is limited to UV-crosslinked polymers. The materials selection, however, broadens each year, new resins with enhanced strength or flexibility are available on the market.

One of the most favorable advantages of SLA & DLP 3D printers is the high accuracy in the produced objects characterized by very smooth surface finishes. This makes them especially famous among artists, for manufacturing sculptures, jewelry molds, and other prototypes. On the other hand, the SLA-DLP technologies are not suitable for printing relatively large or high strength objects. The technology has been accessed as a useful tool in biomedical engineering too [11] (**Figure 4**).

3.3. Selective laser sintering (SLS)

In the process called selective laser sintering (SLS), a high-power laser is required. The laser is employed in order to melt and solidify layers of powder and produce, again layer-by-layer, 3D objects. The SLS printers are commonly equipped with two plates called pistons. First, a first layer of powder is laid onto the fabrication piston. The high-power laser maps/scans the first layer in the powder, thus selectively melting and sintering—the powder material [12]. In this way, the first layer is fabricated. After solidification of the first layer, the fabrication piston is slightly lowered, and the powder delivery bed, in which the power is contained, is raised by some microns. Then, a roller forces another layer of powder on top of the previous solidified layer. The aforementioned procedure is repeated, allowing the laser to melt and solidify all successive layers one by one, until the designed part has been finished bottom to top (**Figure 5**).

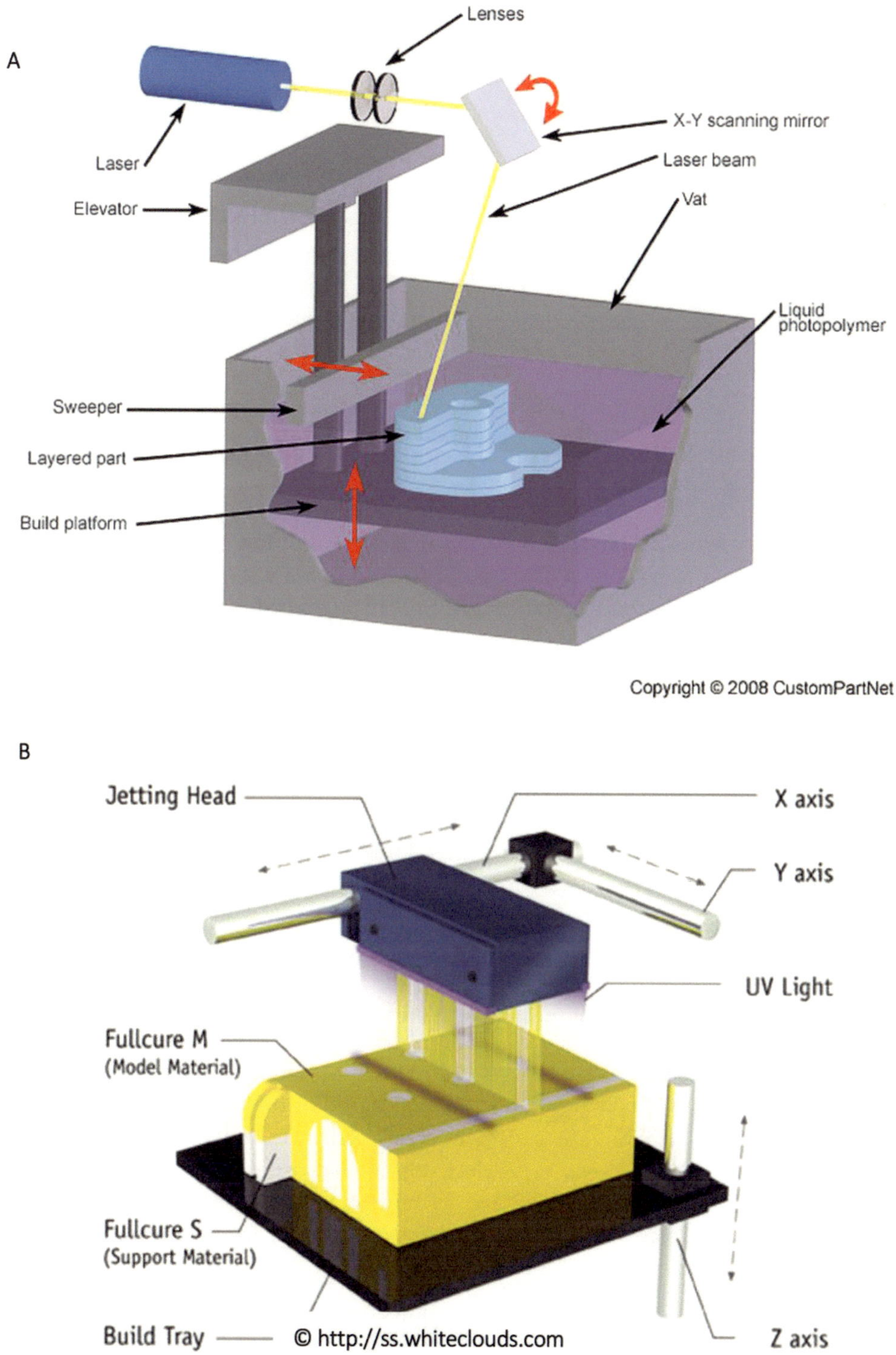

Figure 4. (A) Stereolithography (SLA) vs. (B) digital light processing (DLP) techniques.

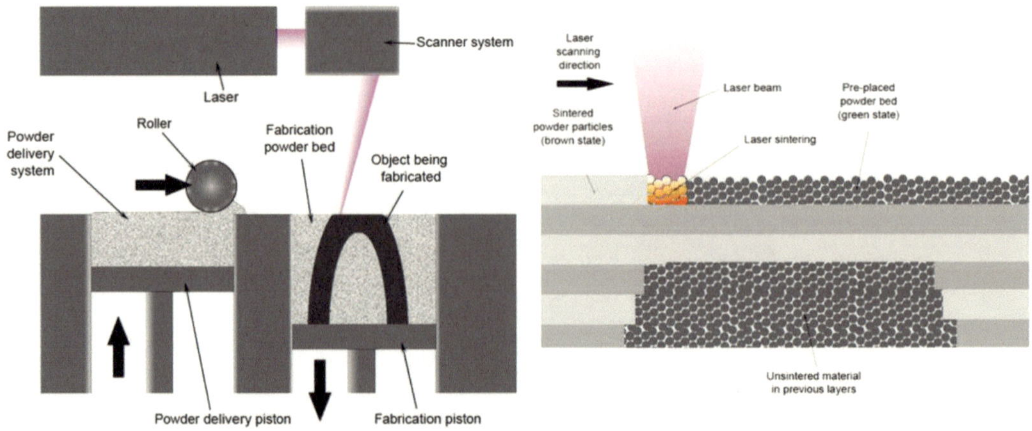

Figure 5. Selective laser sintering (SLS) method. © Materialgeeza—own work, CC BY-SA 3.0, https://commons.wikimedia.org/w/index.php?curid=4032088.

SLS is a highly efficient method, though rather expensive, but has established itself into the industrial 3D printing applications. Desktop SLS printers are widely available on the open market, and prices are already quite affordable. In 2017, a small SLS system could be acquired for as much as $15,000. The usual materials available nowadays as powders for SLS include most thermoplastics such as polyamides (nylon), polystyrene, thermoplastic elastomers, etc.

Due to its high accuracy and production fidelity, SLS machines are widely used for manufacturing end products as well as functional prototypes. Complete design freedom is its most important advantage. There is no need for support of structure. The surrounding unmolten powder acts as a support for the structure as it is layered, which allows for complex, sophisticated, and delicate shapes to be manufactured. Finished objects, as a side effect, take a bit more time to cool, and thus longer lead times are expected. Excess and attached object powder is removed by blowing air or using high pressure water or liquids, and is recycled after filtering; hence, economy of raw materials is *in situ* achieved.

3.4. Metal printing (selective laser melting and electron beam melting)

Selective laser melting and electron beam melting (SLM and EBM) are two of the most common metal 3D printing technologies. They are considered as offspring of the SLS technique described above. Just like SLS, these processes create objects from thin layers material. Raw material is the form of powder and it is selectively melted using an intense heat source. As metals and also ceramics, are characterized by higher melting points, consequently much more power is required; this is provided by a high-power laser for SLM or even an electron beam in the EBM technology.

The printing process begins by distribution of a thin layer of metal powder onto a build plate. The powder is selectively melted by a laser (SLM) [13] or an electron beam (EBM) [14] which maps the object layer. The platform or build plate is afterwards lowered by some microns and rapidly coated with new layer of metal powder on top of the solidified layer. The process is repeated until all layers have been solidified resulting in the finished part. Contrary to SLS, the SLM and EBM techniques require support structures, in order to stabilize the object to the build platform and enable manufacturing of overhanging parts. As a side effect, these enable heat transfer away from the solidified powder. Moreover, SLM is performed in a low-oxygen environment and EBM in vacuum. These conditions enable thermal stresses reduction and warping prevention, and they also allow reactive metals and alloys to be used as raw material.

Due to their high accuracy and costs, SLM and EBM are mainly applied in industrial 3D printing. Materials include various metals and alloys including steel, titanium, aluminum, cobalt-chrome, and nickel.

Metal printing is considered to be as the "holy grail" of additive manufacturing and 3D printing; it has found its path in the aerospace, aircraft, automotive, and healthcare industries for a range of high-tech, low-volume part production, from prototyping to final production. 3D printed metal parts allow for monolithic structuring (reducing the quantity of components), miniaturization, and mass reduction combined with design optimization, as shown in **Figure 6**. SLM and EBM have evolved to a stage where these prints are directly comparable to traditionally manufactured parts in terms of chemical composition, mechanical properties (static and fatigue), as well as microstructure. In the year 2017, for the first time, direct metal laser sintering (DMLS) devices, as the latest generation of SLM printers are referred to, were presented in the market at a cost lower than $100,000. Prices for such systems in the previous years were over half and near 1 M$.

Figure 6. A stainless-steel bracket optimized for weight reduction (front) and the traditional cast bracket in the back. Source: European Space Agency events via Flickr.

4. Advanced applications of the technology of 3D manufacturing

4.1. Automotive industry

3D printing is becoming more and more a familiar technology to the automotive industry, enabling manufacturing of not only prototypes but also finished parts as well. In February 2017, BMW iVentures, the automaker's venture capital arm, announced an investment in Desktop Metal, a startup devoted to 3D printing metal objects. BMW wants to help accelerate the rollout of this technology in both its design and manufacturing departments. In Formula 1, quite a few racing teams have been testing and ultimately creating custom car parts, using 3D printing for prototyping that are used in high speed races. In the same spirit, Swedish car manufacturer Koenigsegg employed 3D printing to manufacture the variable turbocharger for their One:1 model—a car that has an astonishing 1:1 HP-to-Kg curb weight ratio. Although the 100% metal part is not only very lightweight, more importantly, it can also endure the high forces of supercar combustion and demanding racetrack conditions. Other high-tech examples include Ai Design company's bespoke interior items for high-end cars with popular items including housings for radar detectors, iPhones, and aftermarket SatNav units that blend in with the car's interior. The company often services customers with Lamborghinis, Ferraris, and classic Bentleys, so the fit and finish on the OEM-grade thermoplastics has to be perfect. Ai's experienced engineering was reluctant to abandon the highly efficient CNC manufacturing behind. It took a lot of expert consulting from Stratasys to make Ai Design's people to comprehend the potential in manufacturing its models with a 3D printer and fused deposition modeling.

BMW, as mentioned above, went for 3D printing technologies quite early. The company has its own Rapid Manufacturing Facility at the HQ in Munich. BMW is considered to be of the founding fathers of stereolithography having recently revealed plans and approach for a fully 3D printed car. One can understand how deeply 3D printing has become incorporated into the company culture by the fact that even a thumb cast for assembly line workers was produced in that way. Evidently, the workers have to push by thumb, a huge number of rubber plugs into chassis holes on the assembly line. This repetitive work causes a repetitive strain type injury in many of them. Confronting the issue, BMW engineers came up with a bright idea: a cast of the thumb and hand that relieved all the strain out the process. Quite simple, very brilliant, and it just proves just how deep 3D Printing has gone into the corporate culture at BMW. It also shows that 3D printing goes beyond the actual manufacturing process itself into biomechanics and ergonomic concepts (**Figure** 7) [15, 16].

4.2. Medical and dental industry

Being always at the cutting edge of technology, biomedical and prosthetics fields has largely benefited from the introduction of 3D printing in these sectors. Custom-shaped personalized-hearing aids no longer require manual labor to manufacture; with 3D printing, they can be made with the click of a button in a very short time. This of course implies substantially lower costs and shorter production times. Even orthopedic implants manufacturing at custom dimensions from CT or MRI scans from the patients is nowadays feasible.

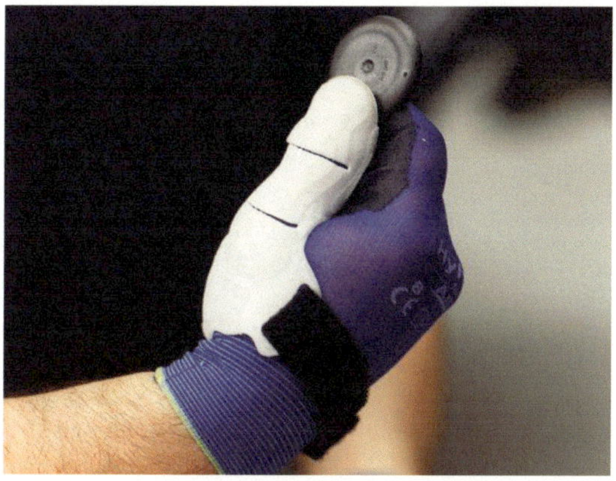

Figure 7. BMW has turned to 3D printing to augment its workers and stop strain on limbs frequently found on manufacturing lines. Photograph: BMW (republished from the Guardian).

Prosthetics and other assistive medical devices, braces, and retainers are tailored specifically for the needs of the patient. This has totally reversed an inherent problem that of time and energy required to manually produce each product. As a natural consequence, introduction of 3D printing in the dental and orthodontics fields was an inevitable event. With today's technology, a dental surgeon or orthodontist can use an intraoral 3D camera to scan a client's oral cavity and teeth, use afterward a specialized software and digitally design dental prosthetics, braces, crowns, bridges, etc. Then, he can send the files to a dental technician to 3D print the required molds or directly print the prosthetic itself. As if it was meant to be invented for them the dental industry fully adopted 3D printing technologies. Nowadays, there are dedicated 3D printer models produced specifically for manufacturing dental aids and molds. Alone 3D printer company Stratasys offers two wax 3D printers available to the dental industry. The Stratasys CrownWorx and FrameWorx 3D Printers are supposed to provide the highest precision in wax 3D printing, allowing dental laboratories to produce wax-ups for crowns, bridges, and denture frameworks. Imagine economy in time and costly silicon imprinting materials and gypsum molding. 3D Printers for dental applications such as Stratasys CrownWorx and FrameWorx use wax deposition modeling (WDM) technology. Mainly based on jetting technology and waxy polymers, they allow production of wax-ups characterized by smooth surface finishes and minimal post-processing effort and time requirements. Stratasys claims that the waxy materials burn out leaving no residue, no material shrinkage, neither invoking cracking, nor expansion (**Figure 8**).

Nowadays, many types of 3D printers are used also in other areas of biomedical applications, such as manufacturing scaffolds for tissue engineering [17, 18] and many other areas of biomaterials engineering [19].

4.3. Aerospace

SpaceX designed and built its famous SuperDraco hypergolic propellant liquid rocket engine. It is a member of SpaceX's Draco rocket engines family. Dragon V2 passenger-carrying space

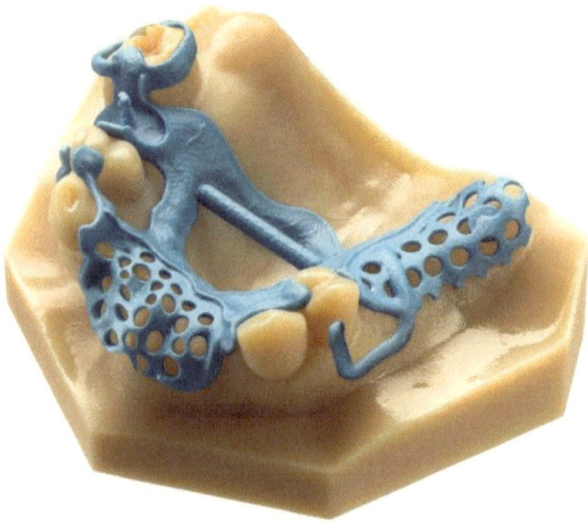

Figure 8. An example of a dental frame built using wax deposition modeling. Source: www.Stratasys.com.

capsule shall be powered by a redundant array of eight SuperDraco engines. These provide fault-tolerant propulsion in the launch escape system and propulsive-landing thrust (**Figure 9**).

The combustion chamber of the SuperDraco space engine is created with direct metal laser sintering (DMLS), using Inconel powder. This super-strong nickel-chromium-based "superalloy" is quite difficult to machine in the traditional way with CNC's. The use of 3D printing DMLS technology "*resulted in an order of magnitude reduction in lead-time compared with traditional machining – the path from the initial concept to the first hotfire was just over three months,*" according to the company's website. Moreover, the combustion chamber is regeneratively cooled. This method allows cryogenic propellant to pass through a jacket covering the combustion chamber, cooling thus the engine, a trusted solution in the rocker motor design technology.

Figure 9. SPACEX, Superdraco engine [20].

On the other hand, in early 2016, rocket and missile propulsion manufacturer Aerojet Rocketdyne, a renowned aerospace and defense leader, was the recipient of a $6 million contract from the US Air Force to define 3D-printed rocket engine component standards. The standards will be used to qualify the 3D printed components used in liquid-fueled rocket engine applications, in order to follow through with a mandate set down by US Congress: that the Department of Defense will stop using Russian-made RD-180 engines to launch US satellites and national security payloads into space and begin using domestically produced options instead. Shortly after, Aerojet signed its own contract with Sigma Labs, to non-exclusively license its PrintRite 3D software system to evaluate and redefine the 3D-printed components used in Air Force manufacturing (**Figure 10**).

Aerojet, being certainly the right company for the Air Force contract, had already successfully completed hot-fire testing of the 3D printed rocket engine injectors for its liquid-fueled AR1 booster rocket engine in 2015; selective laser melting was used to manufacture the components. It has been known that Aerojet has also successfully completed its Critical Design Review (CDR) for the 500,000 foot-pound thrust-class AR1 engine. This achievement will keep the AR1 on track for flight certification in 2019, as a replacement for the Russian RD-180 engine. Twenty-two incremental CDRs came before the recent system-level CDR, along with full-scale testing of critical subsystem components, like the staged combustion system.

In the aviation sector, GE Aviation and Safran companies have successfully launched a method for 3D printing of jet engine fuel nozzles. The remarkable technology allows engineers to replace complex assemblies with a single part. The 3D-printed nozzles are lighter than previous designs, and boost a jet engine's fuel efficiency by up to 15%. LEAP engines of GE equipped with 3D-printed fuel nozzles which will power new generation narrow-body planes, e.g., Boeing 737MAX and Airbus A320neo. Especially, the A320neo Airbus passenger airplane is powered by twin LEAP jet engines with 3D-printed parts based on new advanced materials. LEAP ("Leading Edge Aviation Propulsion") is a high-bypass turbofan engine which, due to its advanced space age materials operates at higher pressures than the previous CFM56 machine. The LEAP is the first engine equipped with actually 19 (!) 3D-printed fuel nozzles and parts

Figure 10. The Aerojet Rocketdyne AR1 booster engine can be configured as a single engine or a twin booster.

from space age, super-strong ceramics that make it 15% more fuel efficient than the previous CFM56 airplane jet engines built by CFM International. CFM is the 50/50 joint-venture between GE Aviation and France's Safran (Snecma) who designed the engine. Airbus picked the LEAP for the A320neo in 2010. Since then, CFM has received more than 2500 orders and commitments for the LEAP-1A engine, representing 55% of A320neo orders to-date (**Figures 11** and **12**).

In August 2016, the LEAP engine was installed on the Airbus A320neo with Pegasus Airlines and CFM delivered 77 machines. Following the introduction of Boeing 737 MAX, CFM delivered 257 LEAPS in the first 9 months of 2017, including 110 in the last three: 49 to Airbus and 61 to Boeing, and targets 450 in the year. The predictions for production at CFM are 1200 engines in 2018, 1900 in 2019, and 2100 in 2020, respectively. This is compared to the 1700 CFM56 produced in 2016. FAA has recently certified the first 3D-printed part for a GE jet engine—a casing that houses the compressor inlet temperature sensor inside the GE90 jet engine.

Figure 11. An Airbus A320neo powered by a pair of LEAP-1A engines took a maiden flight on May 19 in Toulouse, France. The two engines used for the four-and-a-half-hour flight were the LEAP-1A, developed specifically for the Airbus jet [21].

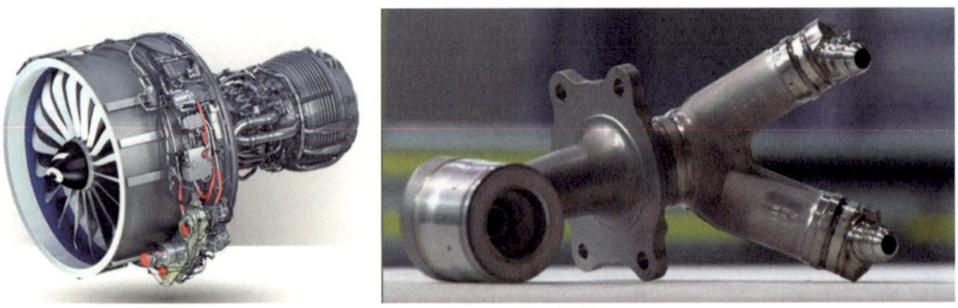

Figure 12. The LEAP engine (to left) has 19 3D-printed fuel nozzles (top right) and static turbine shrouds made from ceramic matrix composites (CMCs) (above left). Image credit: CFM.

http://dx.doi.org/10.5772/intechopen.74331

4.4. Fuel tanks

In the totally innovative additive manufacturing process technology invented by Sciaky [22, 23], it is possible to build the two hemispherical halves of fuel tanks. Layer-by-layer, spools of titanium wire is spun, providing the material which melts and deposits, forming thus the tank walls. Lockheed Martin Space Systems, after carefully reviewing and applying the process, stated that they plan to re-think the way they produce satellite propellant tanks. Moreover, eventually the construction of those tanks will be shifted in-house, and thus results in significant capital savings. Up to this point, Lockheed Martin bought those critical titanium tanks from Orbital ATK. The three Mars orbiters use the Orbital ATK tanks as will NASA's OSIRIS-Rex asteroid-probing space vehicle upon its completion and launch. Lockheed Martin official Ambrose stated that it has become critical to reduce the lead times for building satellites, and he said that 3D printing holds the keys to reaching that goal (**Figure 13**).

Sciaky Inc. had already begun developing the wire-feed electron beam process back in the mid-1960s [24]. The process was further developed in the 1990s allowing for production of jet engine knife edge seals [25]. The EBAM process was more advanced early in the 2000s, allowing manufacturers' significant savings in time and money on the production of large, high-value metal parts. Sciaky formally launched the EBAM process, in 2009, which was then marketed as Electron Beam Direct Manufacturing, as a service option. Lockheed Martin Aeronautics selected Sciaky for the Department of Defense (DOD) Mentor-Protégé Program in 2011. The special focus of this agreement was the application of additive manufacturing for titanium structural components for Lockheed Martin's F-35 Lightning II fighter project. Further on, in 2012, Sciaky signed a partnership with Penn State University, via DARPA (Defense Advanced Research Projects Agency) funding. Their goal was to advance Direct Digital Manufacturing (DDM) technologies for highly engineered and critical metallic systems and components in applications for the Department of Defense (DOD) and US industry.

Finally, in 2014, Sciaky started delivering fully operational commercially available EBAM systems. Needless to mention, Lockheed Martin Space Systems was one of the first customers to acquire an EBAM system for developing and producing "3D-printed" titanium propellant tanks. The EBAM system which Lockheed bought last year from Sciaky Inc. costs $4 million and is capable of "printing" out fuel tanks of nearly 150 cm in diameter. There is even better news: the method cuts down the cost of manufacturing propellant tanks by as much as 50%.

Figure 13. The manufacturing procedure of one half of a satellite fuel tank according to the Sciaky method (top left) and finished half-product (top right). Source: Sciaky Inc. Video capture.

The process is also much faster than casting those tanks in molds. With around 20 months being the lead time of the casting technique, and the time spent procuring the bulk, forged titanium billets also considered, EBAM additive manufacturing is considerably quicker. Material costs are also reduced with respect to structural titanium parts that are machined from a billet or forged. The EBAM spare also all the required involved machining time by as much as 80%. Dennis Little, Lockheed's vice president of production for space systems, stated that, those manufacturing advantages involved in the 3D EBAM-printed titanium tanks, will be in use on spacecraft before the decade is out, provided that internal evaluation of the process meets certification requirements from the US Air Force and NASA.

4.5. 3D printed drones

Drone fashion hobby, also used in serious applications, has dramatically increased in the past several years. Global unit sales grew 60% (to 2.2 million) and revenue increased 36% (to $4.5 billion) in 2016 [26]. American consumers in the United States alone, bought 2.4 million hobbyist drones, compared to 1.1 million in the previous year. In 2017, revenue is anticipated to reach $6 billion, while units should rise up to 3 million. The latter figures substantiate a revenue growth of 39% and unit growth 34% within 2 years time, according to a study conducted by the research firm Gartner Inc. [27].

Impressively, engineers, designers, and other individual end-users become more and more involved in current research and development efforts to manufacture 3D-printed drones are now eligible for R&D federal and state tax credits.

Military sectors are intensively exploring novel paths to make cheaper, lighter, and more energy-efficient drones. A Marine Corp named Rhet McNeal created Scout, a drone composed of 3D-printed components. Scout only costs $600 to build. In comparison, a similar military-grade drone costs hundreds of thousands of dollars to build from advanced materials with conventional methods. Another benefit which arises from the technology, being a 3D-printed drone, should it receive any damage, any part or parts can be directly printed out and installed within hours. On the other side of the fence, a standard-issue drone would require weeks, sometimes months, to get a replacement through the Marine Corps' supply chain. Scout has been delivered to Mitre Corp., a USMC drone supplier, for certification testing. Prior to Scout, Mitre Corp. experimented with a 3D-printed drone called Nibbler. In the following period of time, the USMC is planning to test Nibbler into a real combat zone in order to supply troops with required resources. At the same time, they are investing resources in R&D on how to manufacture 3D-printed drones for surveillance purposes.

The University of Virginia, in its term, designed and created a 3D-printed UAV drone for the Department of Defense. The Drone can be printed in less than 24 h at an end-user price of $2500, electronics development included. The fuselage of the drone costs only $800. Resembling to one long wing, it was nicknamed the Razor. The Razor can fly at maximum speed of 40 mph for up to 45 min, having a gross weight at 6 pounds with all the equipment installed.

None of its features and capabilities is compromised by the fact that it is 3D-printed; in fact, it features all the same functions and operational capabilities as a typical military-grade drone, including GPS waypoints for navigation and mile-distance control. Even camera hoisting and phone linking capabilities that extend the distance it can be controlled are present. Being 3D printed, it has even greater advantage that it can be modified and reprinted by desire and need. It is fully customizable, as it can be made smaller or bigger, geared to carry a sensor instead of a camera, or fly slower or faster as each different operation requires.

Solid Concepts uses additive manufacturing to produce fixed-wing UAS airframes (such as the PTERA shown in **Figure 14**) that are used to test high-risk circulation control systems, conformal fuel tank concepts, and other advanced aerospace concepts with SLS 3D-printed parts.

4.6. 4D printing

Additive manufacturing—or 3D printing—is almost 30 years old. Today, it is not only just found in industry but also in households, as the price of high accuracy 3D printers has fallen below US$1000. Understanding of this new power enabling us to design and print almost anything in three dimensions, not just scribes and symbols on paper, opens up unlimited opportunities for everyday people to manufacture from toys and household appliances, to jewels and tools, in our homes and work places.

Well, guess again, there is even more that can be done with 3D-printed materials. We can make them more flexible and more useful. Smart structures can be also printed, featuring embedded sensors and actuators. These objects, also 3D-printed, can transform in a pre-programmed way in response to an external stimulus. Such a technology enabling objects, parts, or even complete systems to be manufactured was baptized by the popular science name of "4D printing." Perhaps, it leads us to a better or easier way to think about that the object transforms or reacts over time.

Of course, such a behavior of structural deformations is not at all quite new. Researchers have been for decades researching and demonstrating "memory" and "smart material" effects and properties. In their recent Nature article, Raviv et al. propose a new design of complex self-evolving structures that vary over time due to environmental interactions [28]. Logically, in conventional 3D printing systems, materials are expected to exhibit a stable response rather than an active one and fabricated objects are designed and printed as static items or functional parts at the most. In this paper, a novel approach is introduced for simulating and fabricating self-evolving structures. These "smart" structures transform

Figure 14. 3D-printed military drone-PTERA of solid concepts.

into a predetermined shape, by changing properties and functions after fabrication has been completed. The new locally coordinated bending primitives combine into a single system, allowing for a global deformation which can stretch, fold, and bend in the given environmental stimulus. The physics of 4D printing often requires multiple materials to be embedded into a single 3D structure. Much work is anticipated to be invested on such 4D printers and structures over the next decades.

5. Conclusions

In our days, it is apparent that we experience a very large revolution as far as novel production techniques is concerned, in manufacturing structures by advanced layering techniques, processes widely known under the simplistic name of 3D printing, for the wide public. The truth is that this new kind of industrial revolution regarding manufacturing has not been completed yet. Although techniques for glass and ceramics similar to 3D printing are developed as these lines are written, no mature methods have been proposed yet. Moreover, upon completion of this revolution, which owes a lot to electronics, software (CAD), and computer technology, we will be able to manufacture almost all our products utilizing 3D printing or even 4D printing technologies for smart structures.

It is of course common sense that this will influence our way of thinking and designing items and parts or even whole devices at once, since results of simulation and design optimization are nowadays directly printable. Mankind will benefit, as already does, from applications ranging from medical equipment and implants, biomaterials, or biomedical devices, to possible automated orbital factories fully equipped with 3D printers for space exploration equipment. The limits for this new technology for manufacturing are far from being set as yet. Transport and automotive sectors are also supposed to profit from the impact of 3D printing technologies.

Since these techniques are saving energy and CO_2 emissions, and their products or by-products are recyclable too; it is also a revolution of green manufacturing. Furthermore, as scientists are moving into MEMS and NEMS device manufacturing techniques, it is certainly implied that in some future era, mankind will possess technology to perform atom by atom structuring. This has already been shown. From atom scale to nano-, micro-, and macroscale, we are almost able to manipulate matter totally.

Author details

Dionysios E. Mouzakis

Address all correspondence to: demouzakis@sse.gr

Department of Military Sciences, Hellenic Army Academy, Sector of Mathematics and Engineering Applications, Greece

References

[1] Wohlers T, Gorne T. History of Additive Manufacturing. Wohlers Report 2014. Colorado, USA: Wohlers Associates Inc.; 2014. ISBN: 978-0-9913332-0-2

[2] Kodama H. A scheme for three-dimensional display by automatic fabrication of three-dimensional model. IECE. 1981;**J64-C**(4):237-241

[3] Kodama H. Automatic method for fabricating a three-dimensional plastic model with photo-hardening polymer. Review of Scientific Instruments. 1981;**52**(11):1770-1773

[4] Amir Bashir. Thoughts on Startups. Market Size: How Large is the Market for Consumer 3D Printers? [Internet]. 2014. Available from: http://amirsthoughts.com/market-size-how-large-is-the-market-for-consumer-3d-printers/ [Accessed: Oct 30, 2017]

[5] Caliskan OU, Durgun I. Silicon moulding production technology applications in automotive product development process. In: Proceedings of the 12th Automotive and Production Technologies Symposium, 2011, Bursa, Turkey; Available from: https://www.researchgate.net/publication/280052438_SILICON_MOULDING_PRODUCTION_TECHNOLOGY_APPLICATIONS_IN_AUTOMOTIVE_PRODUCT_DEVELOPMENT_PROCESS [Accessed: Oct 30, 2017]

[6] Durgun I. Sheet metal forming using FDM rapid prototype tool. Rapid Prototyping Journal. 2015;**21**(4):412-422

[7] Rayna T, Striukova L. From rapid prototyping to home fabrication: How 3D printing is changing business model innovation. Technological Forecasting and Social Change. 2016;**102**:214-224. ISSN 0040-1625. DOI: 10.1016/j.techfore.2015.07.023

[8] Szykiedans K, Credo W. Mechanical properties of FDM and SLA low-cost 3-D prints. Procedia Engineering. 2016;**136**:257-262. ISSN 1877-7058. DOI: 10.1016/j.proeng.2016.01.207

[9] Bayraktar Ö, Uzun G, Çakiroğlu R, Guldas A. Experimental study on the 3D-printed plastic parts and predicting the mechanical properties using artificial neural networks. Polymers for Advanced Technologies, 2017;**28**:1044-1051. DOI: 10.1002/pat.3960

[10] Jacobs PF. Stereolithography and other RP&M technologies: From rapid prototyping to rapid tooling. Jointly published by Society of Manufacturing Engineers in cooperation with the Rapid Prototyping Association of SME. Dearborn, Michigan, New York: ASME Press, American Society of Mechanical Engineers; 1995. ISBN: 0872634671

[11] Ferry PW, Melchels JF, Grijpma DW. A review on stereolithography and its applications in biomedical engineering. Biomaterials. 2010;**31**(24):6121-6130. ISSN 0142-9612. DOI: 10.1016/j.biomaterials.2010.04.050

[12] Kruth J-P, Mercelis P, Van Vaerenbergh J, Froyen L, Rombouts M. Binding mechanisms in selective laser sintering and selective laser melting. Rapid Prototyping Journal. 2005;**11**(1):26-36. DOI: 10.1108/13552540510573365

[13] Agarwala M, Bourell D, Beaman J, Marcus H, Barlow J. Direct selective laser sintering of metals. Rapid Prototyping Journal. 1995;**1**(1):26-36. DOI: 10.1108/13552549510078113

[14] Murr LE, Gaytan SM, Ramirez DA, Martinez E, Hernandez J, Amato KN, Shindo PW, Medina FR, Wicker RB. Metal fabrication by additive manufacturing using laser and electron beam melting technologies. Journal of Materials Science & Technology. 2012; **28**(1):1-14. ISSN 1005-0302. DOI: 10.1016/S1005-0302(12)60016-4

[15] Zoran A. The 3D printed flute: Digital fabrication and design of musical instruments. Journal of New Music Research. 2011;**40**(4):379-387. DOI: 10.1080/09298215.2011.621541

[16] Lopez SM, Wright PK. The role of rapid prototyping in the product development process: A case study on the ergonomic factors of handheld video games. Rapid Prototyping Journal. 2002;**8**(2):116-125. DOI: 10.1108/13552540210420989

[17] Mohanty S, Larsen LB, Trifol J, Szabo P, Reddy Burri HV, Canali C, Dufva M, Emnéus J, Wolff A. Fabrication of scalable and structured tissue engineering scaffolds using water dissolvable sacrificial 3D printed moulds. Materials Science and Engineering: C. 2015;**55**:569-578. DOI: 10.1016/j.msec.2015.06.002

[18] Bose S, Vahabzadeh S, Bandyopadhyay A. Bone tissue engineering using 3D printing. Materials Today. 2013;**16**(12):496-504. DOI: 10.1016/j.mattod.2013.11.017

[19] Chia HN, Wu BM. Recent advances in 3D printing of biomaterials. Journal of Biological Engineering. 2015;**9**:4. DOI: 10.1186/s13036-015-0001-4

[20] Rae Botsford End. SpaceX's SuperDraco engine: Abort Capability All the Way to Orbit. [Internet]. 2015. Available from: http://www.spaceflightinsider.com/organizations/space-exploration-technologies/spacexs-superdraco-engine/#pQkTSuUXZlmAvJHe.99 [Accessed: Oct 30, 2017]

[21] Kellner T. CFM International, Jet Engines with 3D-Printed Parts Power Next-Gen Airbus Passenger Jet. [Internet]. 2015. Available from: https://www.ge.com/reports/post/119370423770/jet-engines-with-3d-printed-parts-power-next-gen/ [Accessed: Oct 30, 2017]

[22] Lathabai S, Glenn M, Ritch DR. Comparison of microstructural evolution in Ti-6Al-4V multi-layer builds produced by three additive manufacturing processes. In: 25th Advanced Aerospace Materials and Processes (AeroMat); Jun 17, 2014; Orlando, Florida. Available from: https://www.researchgate.net/publication/267901713_Comparison_of_Microstructural_Evolution_in_Ti-6Al-4V_Multi-Layer_Builds_Produced_By_Three_Additive_Manufacturing_Processes [Accessed: Oct 30, 2017]

[23] Ding D, Pan Z, Cuiuri D, Li H. Wire-feed additive manufacturing of metal components: Technologies, developments and future interests. International Journal of Advanced Manufacturing Technology. 2015;**81**(1):465-481

[24] Simon. Wire-Feed Additive Manufacturing Might be the Future of Metal-based 3D Printing. [Internet]. 2015. Available from: http://www.3ders.org/articles/20150531-wire-feed-additive-manufacturing-might-be-the-future-of-metal-based-3d-printing.html [Accessed: Oct 30, 2017]

[25] Recommended Practices for Electron Beam Welding and Allied Processes, Practical Examples 13.2 – Knife Edge Seal Using Electron Beam Additive Manufacturing Process American National Standard. Doral, Florida: American Welding Society; AWS C7.1:2013. ISBN: 978-0-87171-835-8

[26] Goulding C. 3D Printed Drones Fly into the Future. [Internet]. 2017. Available from: https://3dprint.com/185369/3d-printed-drones-tax-credits/ [Accessed: Oct 30, 2017]

[27] Glaser A. DJI is Running Away with the Drone Market, Better Products at Lower Prices are Helping China's Drone Giant Take Control of a Fast-growing Industry. [Internet]. 2017. Available from: https://www.recode.net/2017/4/14/14690576/drone-market-share-growth-charts-dji-forecast [Accessed: Oct 30, 2017]

[28] Raviv D, Zhao W, McKnelly C, Papadopoulou A, Kadambi A, Shi B, Hirsch S, Dikovsky M, Zyracki D, Olguin C, Raskar R, Tibbits S. Active printed materials for complex self-evolving deformations. Scientific Reports. 2014;**4**:7422. DOI: 10.1038/srep07422